Homework Helpers

Multiplication Grade 4

As a parent, you want your child to enjoy learning and to do well in school. The activities in the *Homework Helpers* series will help your child develop the skills and self-confidence that lead to success. Humorous illustrations and diverse formats make the activities interesting for your child.

HOW TO USE THIS BOOK

- Provide a quiet, comfortable place to work with your child.
- Plan a special time to work with your child. Create a warm, accepting atmosphere so your child will enjoy spending this time with you. Limit each session to one or two activities.
- Make sure your child understands the directions before beginning an activity.
- Check the answers with your child as soon as an activity has been completed. (Be sure to remove the answer pages from the center of the book before your child uses the book.)
- The activities in this book were selected from previously published Frank Schaffer materials.
- Topics covered in this book are basic facts 0–12, multiplication by multiples of 10 and 100, missing factors, multiplication of two-digit and three-digit numbers by one-digit numbers, multiplication of two-digit and three-digit numbers by two-digit numbers, regrouping, estimation, multiplying money, and finding factors.

ISBN #0-86734-110-6

FS-8143 Homework Helpers—Multiplication Grade 4

1028 Via Mirabel, Palos Verdes Estates, CA 90274

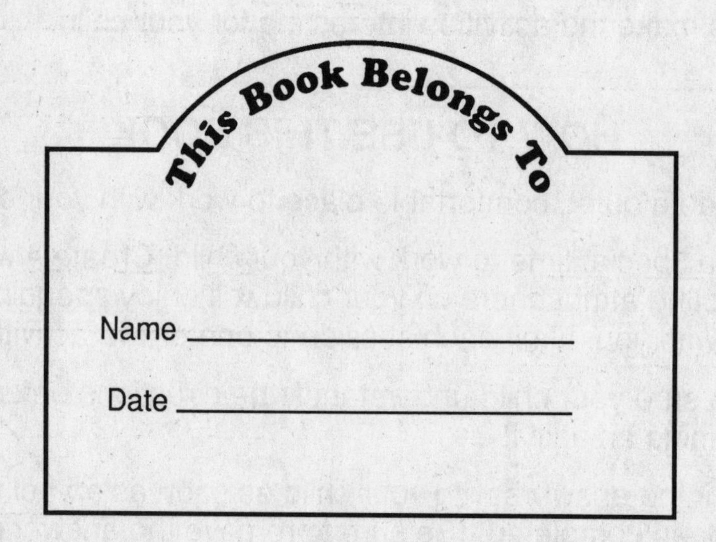
This Book Belongs To
Name
Date

Tennis Match

Multiply and write each product.

A.	0 x 5	0 x 7	0 x 2	0 x 4	0 x 6
	0	0	0	0	0

B.	0 x 0	0 x 9	0 x 3	0 x 1	0 x 8
	0	0	0	0	0

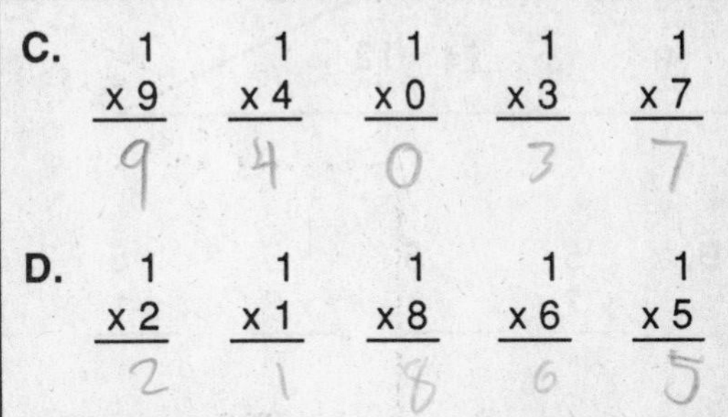

C.	1 x 9	1 x 4	1 x 0	1 x 3	1 x 7
	9	4	0	3	7

D.	1 x 2	1 x 1	1 x 8	1 x 6	1 x 5
	2	1	8	6	5

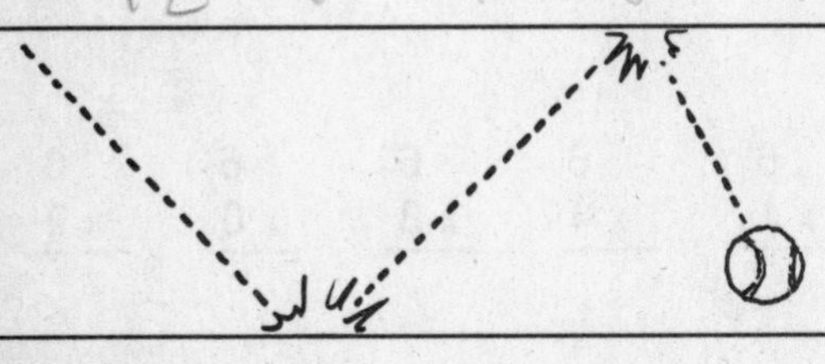

E.	2 x 8	2 x 3	2 x 5	2 x 0	2 x 1
	16	6	10	0	2

F.	2 x 6	2 x 9	2 x 7	2 x 4	2 x 2
	12	18	14	8	4

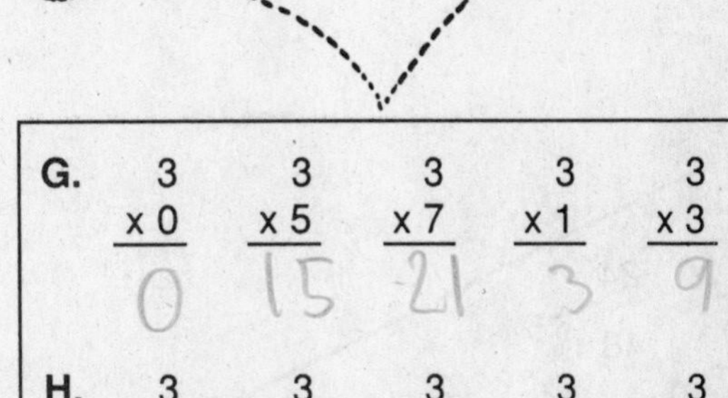

G.	3 x 0	3 x 5	3 x 7	3 x 1	3 x 3
	0	15	21	3	9

H.	3 x 2	3 x 9	3 x 6	3 x 4	3 x 8
	6	27	18	12	24

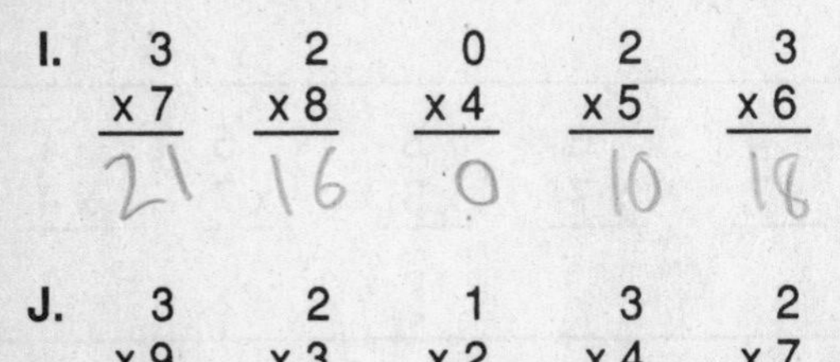

I.	3 x 7	2 x 8	0 x 4	2 x 5	3 x 6
	21	16	0	10	18

J.	3 x 9	2 x 3	1 x 2	3 x 4	2 x 7
	27	6	2	12	14

Brainwork! Matt won four games. Lindsay won twice as many games as Matt. How many games did Lindsay win?

Skateboard Challenge

Multiply and write each product. Cross off the answer on the matching skateboard ramp.

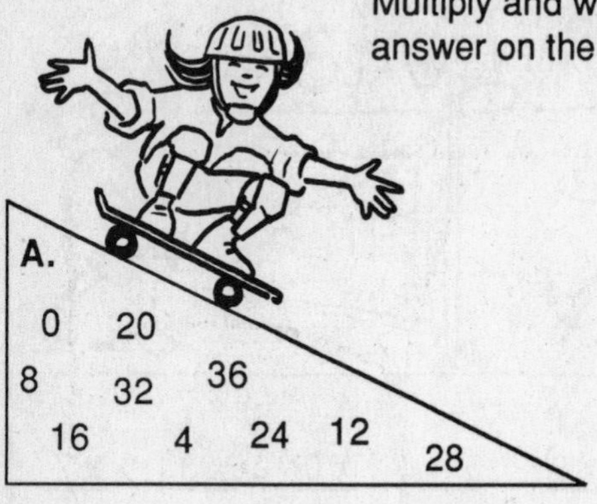

A.

4	4	4	4	4
x 8	x 4	x 2	x 3	x 7
32	16	8	12	28

4	4	4	4	4
x 5	x 1	x 6	x 9	x 0
20	4	24	36	0

B.

5	5	5	5	5
x 7	x 3	x 5	x 1	x 2
35	15	25	5	10

5	5	5	5	5
x 6	x 4	x 0	x 9	x 8
30	20	0	45	40

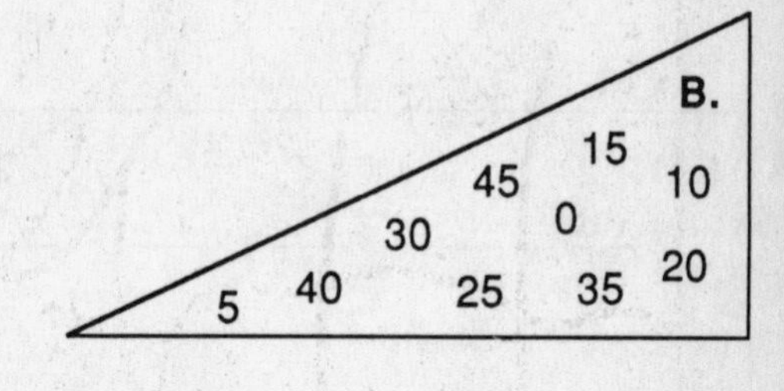

C.

6	6	6	6	6
x 3	x 5	x 2	x 6	x 7
18	30	12	36	42

6	6	6	6	6
x 1	x 4	x 8	x 0	x 9
6	24	48	0	54

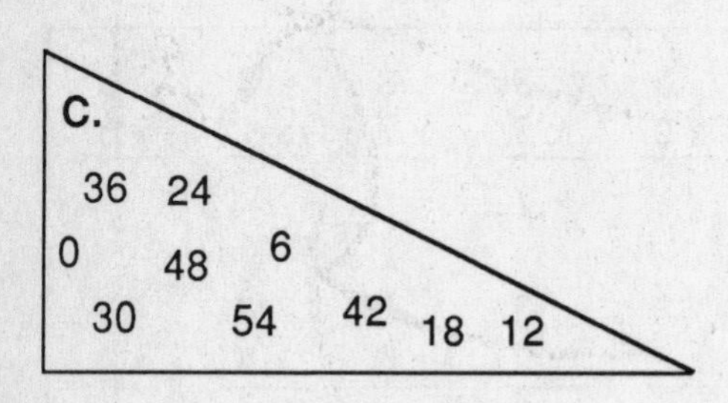

Solve each problem.

D.

4	6	6	4	5	4	6	5	5	4
x 3	x 2	x 9	x 1	x 8	x 7	x 7	x 3	x 7	x 4
12	12	54	4	40	28	45	15	35	16

E.

5	6	4	5	6	4	6	4	5	5
x 2	x 8	x 9	x 1	x 6	x 0	x 3	x 8	x 5	x 6
10	48	36	5	36	0	18	32	25	30

12	18	20	24	30
green	red	orange	brown	yellow

Net Action

Multiply and write each product.
Cross off the answer on the matching net.

A.

7	7	7	7	7
x 7	x 3	x 4	x 1	x 8

7	7	7	7	7
x 5	x 9	x 2	x 0	x 6

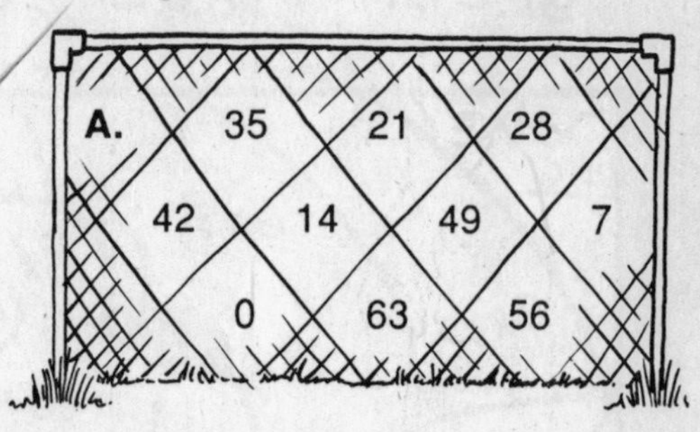

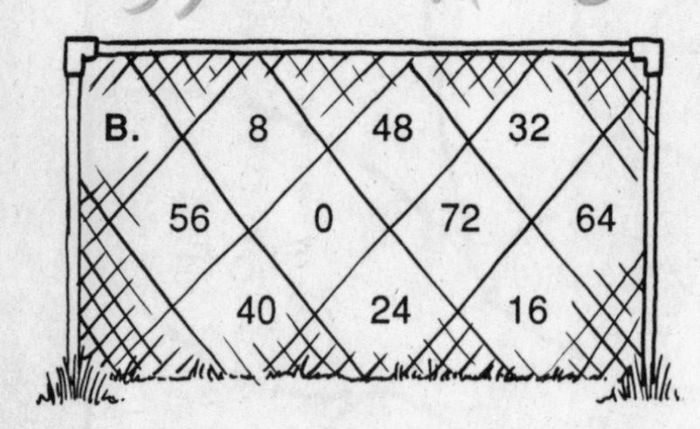

B.

8	8	8	8	8
x 3	x 7	x 5	x 1	x 9

8	8	8	8	8
x 4	x 6	x 8	x 0	x 2

C.

9	9	9	9	9
x 5	x 7	x 3	x 2	x 4

9	9	9	9	9
x 1	x 8	x 9	x 6	x 0

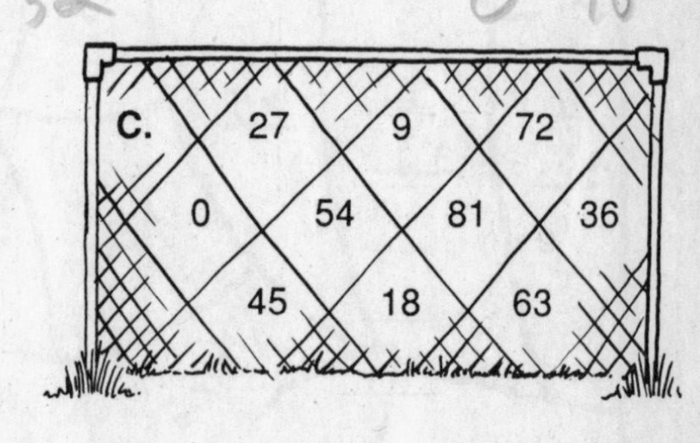

Solve each problem.

D. 7	9	8	8	9	7	9	7	8	9
x 3	x 2	x 6	x 7	x 0	x 6	x 5	x 7	x 4	x 3
E. 8	7	9	8	7	9	7	8	9	8
x 8	x 4	x 6	x 1	x 9	x 9	x 2	x 5	x 4	x 9

Brainwork! Choose five products from the nets. Write one multiplication problem for each.

Bowling for Products

Multiply and write the answers.

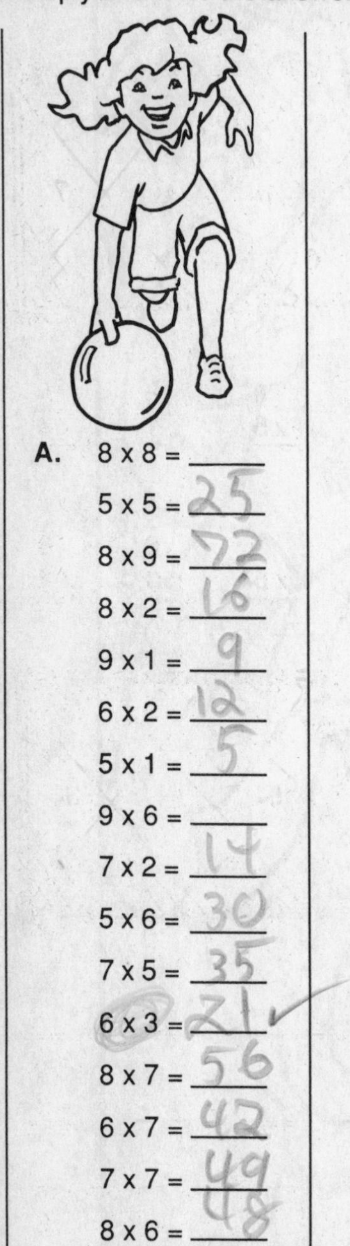

A.

8 x 8 = ____
5 x 5 = 25
8 x 9 = 72
8 x 2 = 16
9 x 1 = 9
6 x 2 = 12
5 x 1 = 5
9 x 6 = ____
7 x 2 = 14
5 x 6 = 30
7 x 5 = 35
6 x 3 = 21
8 x 7 = 56
6 x 7 = 42
7 x 7 = 49
8 x 6 = 48
9 x 8 = 72

B.

6 x 1 = 6
8 x 1 = 8
6 x 9 = ____
9 x 0 = 0
5 x 0 = 0
7 x 1 = 7
7 x 9 = ____
5 x 9 = ____
8 x 3 = ____
9 x 2 = 18
8 x 4 = 32
6 x 0 = 0
5 x 2 = 10
9 x 4 = ____
6 x 5 = 30
7 x 8 = 56
9 x 7 = 63

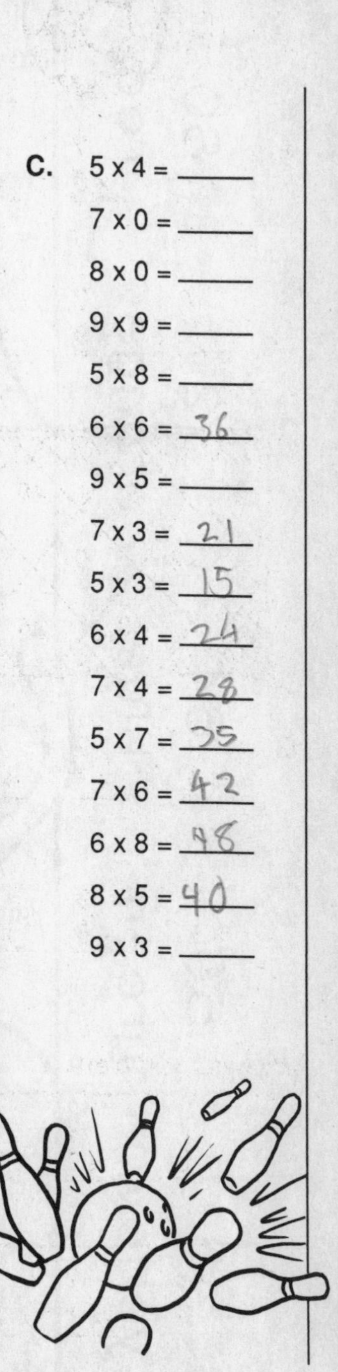

C.

5 x 4 = ____
7 x 0 = ____
8 x 0 = ____
9 x 9 = ____
5 x 8 = ____
6 x 6 = 36
9 x 5 = ____
7 x 3 = 21
5 x 3 = 15
6 x 4 = 24
7 x 4 = 28
5 x 7 = 35
7 x 6 = 42
6 x 8 = 48
8 x 5 = 40
9 x 3 = ____

Brainwork! Choose five multiplication facts you need to practice. Write each one using words in place of numbers and symbols. Example: five times seven equals thirty-five

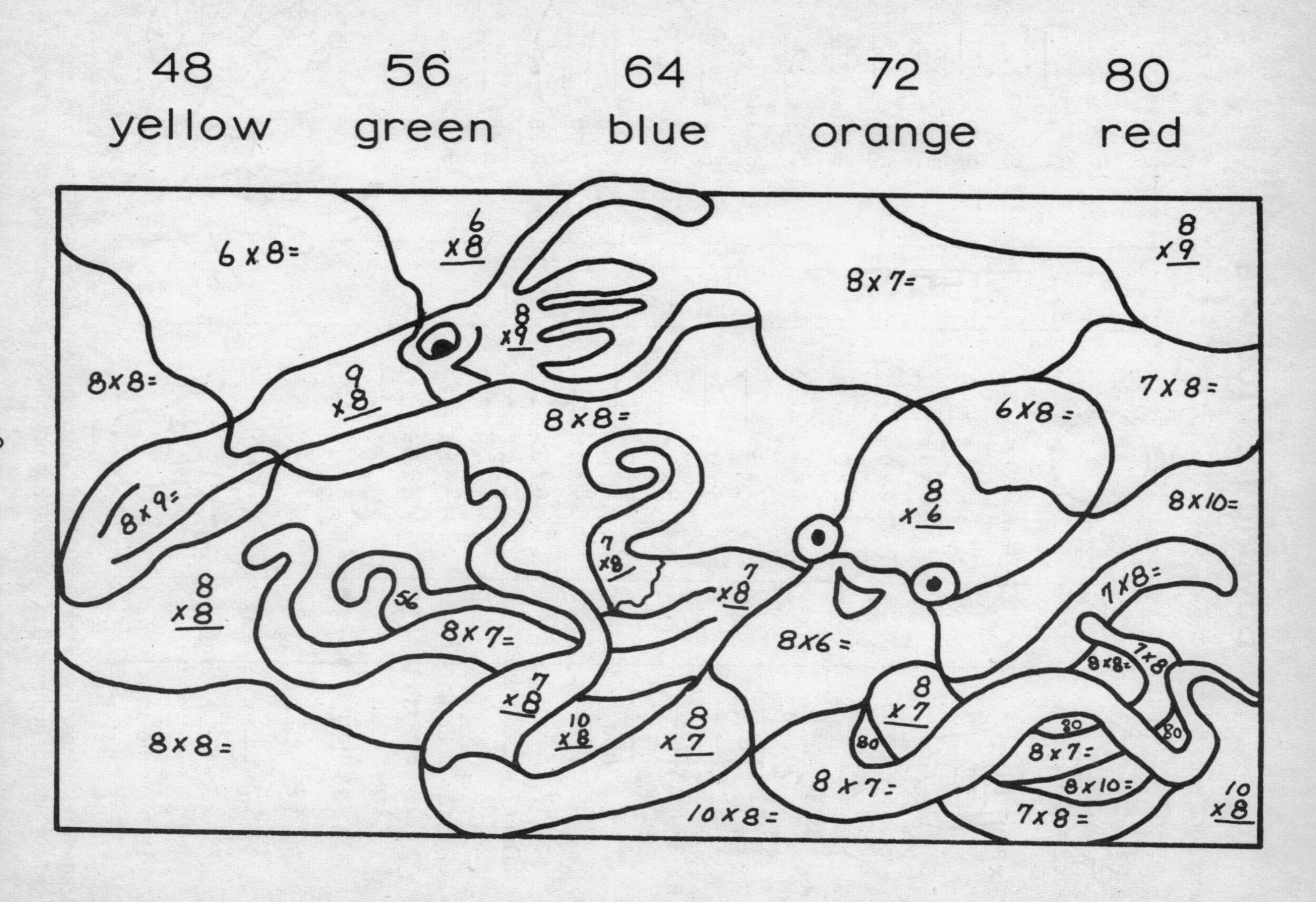
48
yellow
56
green
64
blue
72
orange
80
red

Math Decoder

Be a math decoder! Answer the math problems to find the code. Then use the code to find the answers to the riddles.

2 x 2 =	4	A	3 x 2 =	6	H	5 x 8 =	40	O	8 x 0 =	0	V
3 x 3 =	9	B	4 x 8 =	32	I	7 x 1 =	7	P	9 x 7 =	63	W
4 x 3 =	12	C	5 x 5 =	25	J	8 x 3 =	24	Q	7 x 7 =	49	X
8 x 8 =	64	D	1 x 9 =	9	K	4 x 5 =	20	R	4 x 4 =	16	Y
5 x 3 =	15	E	7 x 5 =	35	L	5 x 6 =	30	S	5 x 9 =	45	Z
7 x 4 =	28	F	2 x 4 =	8	M	6 x 8 =	48	T			
6 x 6 =	36	G	9 x 6 =	54	N	3 x 6 =	18	U			

1. Riddle: What kind of house weighs the least?
   Answer: 4 35-32-36-6-48-6-40-18-30-15

   Decode the answer: LIGHTHOUSE

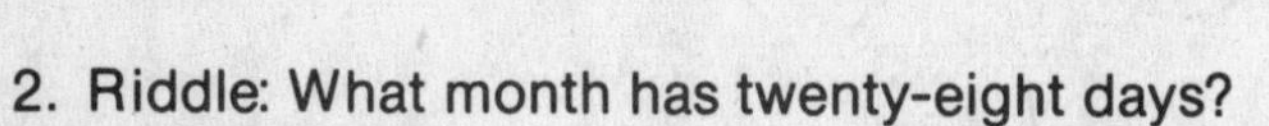

2. Riddle: What month has twenty-eight days?
   Answer: 4-35-35 40-28 48-6-15-8

   Decode the answer: ALL OF THEM

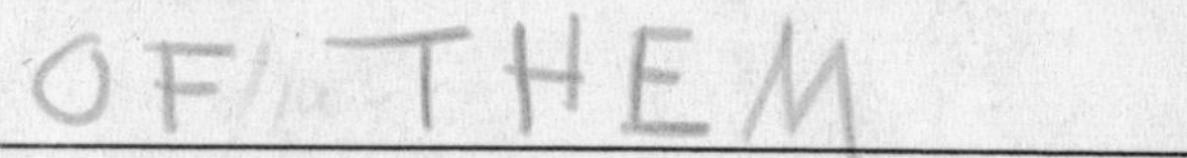

3. Riddle: What did one penny say to the other?
   Answer: 48-40-36-15-48-6-15-20 63-15 8-4-9-15 8-40-20-15 12-15-54-48-30

   Decode the answer: TOGETHER WE MAKE CENTS

Brainwork! Write your name in code.

35-15-15 6-18-54-48

Dribble and Drill

Multiply.

A.	6 x 8	9 x 3	2 x 1	4 x 0	4 x 8	0 x 2	5 x 7	9 x 2	1 x 0	7 x 9
B.	8 x 1	1 x 1	6 x 9	5 x 6	3 x 0	6 x 7	4 x 7	3 x 8	6 x 0	9 x 5
C.	2 x 7	7 x 5	9 x 4	0 x 8	2 x 6	3 x 1	8 x 2	1 x 6	2 x 9	9 x 6
D.	4 x 1	5 x 5	1 x 8	7 x 6	5 x 1	1 x 7	5 x 8	7 x 0	0 x 0	3 x 7
E.	0 x 1	7 x 4	3 x 3	6 x 6	1 x 2	4 x 2	2 x 5	0 x 5	9 x 1	8 x 6
F.	5 x 9	8 x 9	3 x 9	1 x 9	7 x 8	2 x 8	6 x 1	2 x 4	6 x 5	0 x 9
G.	8 x 3	6 x 4	1 x 3	5 x 4	3 x 2	0 x 3	7 x 3	3 x 6	8 x 5	4 x 9
H.	4 x 3	0 x 6	7 x 2	2 x 3	8 x 4	7 x 7	5 x 5	2 x 2	4 x 4	9 x 9
I.	8 x 8	9 x 0	6 x 3	5 x 2	8 x 7	4 x 4	5 x 4	1 x 4	7 x 1	9 x 8
J.	9 x 7	1 x 5	4 x 6	0 x 7	3 x 5	2 x 2	3 x 4	6 x 2	0 x 4	8 x 0

Brainwork! Write a word problem about basketball that can be solved by multiplying.

Over the Hurdle

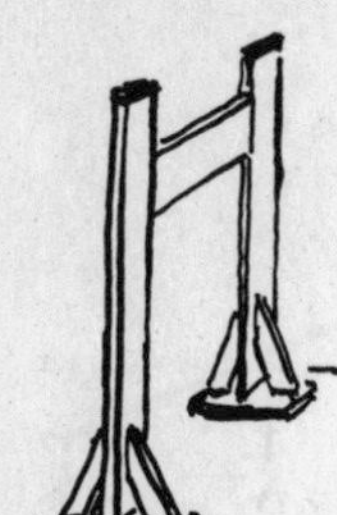

Multiply and write each product.

A.	3 x 4 = ____	6 x 5 = ____	0 x 9 = ____	4 x 5 = ____	1 x 7 = ____
B.	8 x 8 = ____	5 x 4 = ____	7 x 0 = ____	2 x 9 = ____	7 x 8 = ____
C.	2 x 0 = ____	8 x 7 = ____	1 x 8 = ____	6 x 4 = ____	8 x 0 = ____
D.	9 x 5 = ____	0 x 3 = ____	9 x 9 = ____	7 x 9 = ____	3 x 5 = ____
E.	4 x 4 = ____	9 x 6 = ____	9 x 4 = ____	0 x 2 = ____	4 x 6 = ____
F.	1 x 6 = ____	3 x 0 = ____	5 x 3 = ____	8 x 6 = ____	1 x 1 = ____
G.	5 x 5 = ____	6 x 3 = ____	7 x 1 = ____	1 x 9 = ____	5 x 7 = ____
H.	7 x 2 = ____	1 x 0 = ____	4 x 3 = ____	6 x 6 = ____	5 x 2 = ____
I.	0 x 4 = ____	8 x 1 = ____	2 x 1 = ____	2 x 7 = ____	6 x 8 = ____
J.	8 x 2 = ____	3 x 6 = ____	7 x 3 = ____	5 x 1 = ____	3 x 1 = ____
K.	6 x 9 = ____	0 x 6 = ____	4 x 7 = ____	3 x 9 = ____	0 x 5 = ____
L.	2 x 2 = ____	6 x 2 = ____	1 x 5 = ____	5 x 6 = ____	7 x 7 = ____
M.	9 x 3 = ____	2 x 4 = ____	8 x 9 = ____	4 x 2 = ____	2 x 3 = ____
N.	4 x 8 = ____	6 x 7 = ____	0 x 1 = ____	1 x 2 = ____	6 x 1 = ____
O.	7 x 6 = ____	3 x 2 = ____	8 x 3 = ____	9 x 0 = ____	4 x 9 = ____
P.	1 x 3 = ____	0 x 8 = ____	5 x 8 = ____	3 x 7 = ____	3 x 3 = ____
Q.	8 x 5 = ____	9 x 7 = ____	7 x 5 = ____	2 x 5 = ____	9 x 8 = ____
R.	4 x 1 = ____	2 x 8 = ____	5 x 0 = ____	7 x 4 = ____	3 x 8 = ____
S.	5 x 9 = ____	0 x 0 = ____	9 x 1 = ____	1 x 4 = ____	8 x 4 = ____
T.	2 x 6 = ____	9 x 2 = ____	4 x 0 = ____	6 x 0 = ____	0 x 7 = ____

Brainwork! How would you solve this multiplication problem? 4 x 2 x 3 = ________
Write the steps.

Flying High With Math

Fill in the missing numbers in each multiplication chart below.

x	2		8	6			3	1	7	
2	4	10				0				18
	6				12					
4										
							24			

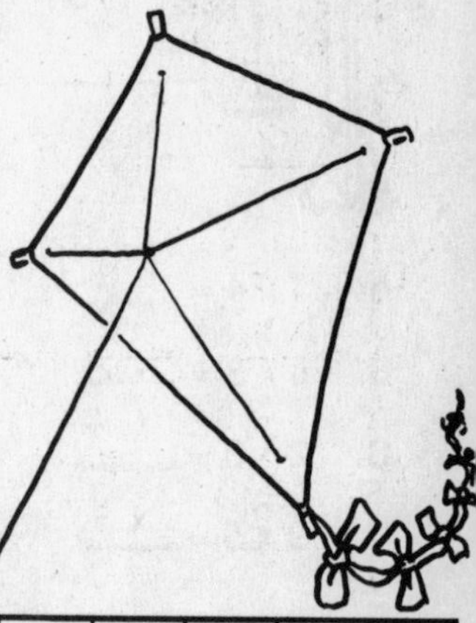

x	6		1
4			
7			
	30		
6		48	
		24	
			2
			9
8			
0			
1			

x	5		7	3		8		6	0	
4		16					36			
			49							7
9					18					

Brainwork! Draw and partially complete a multiplication chart similar to one above. Have a friend fill in the blank boxes. Then check your friend's work.

Ropin' in the Facts

Multiply and write each product.

A.	5 x 2	1 x 9	3 x 5	8 x 3	0 x 9	9 x 10	7 x 6	8 x 4	10 x 2	1 x 8
	10	9	15	24	0	90	42	32	20	8
B.	9 x 9	3 x 6	4 x 8	1 x 7	7 x 7	5 x 3	2 x 1	5 x 9	4 x 7	3 x 6
	81	18	32	7	49	15	2	45	28	16
C.	2 x 8	9 x 6	3 x 4	8 x 2	4 x 0	6 x 5	9 x 7	5 x 4	1 x 6	7 x 5
D.	6 x 3	4 x 9	2 x 0	5 x 1	3 x 7	2 x 9	0 x 6	10 x 0	8 x 5	6 x 4
E.	7 x 4	3 x 3	8 x 7	4 x 1	9 x 4	4 x 6	4 x 10	6 x 2	9 x 8	2 x 3
F.	2 x 7	10 x 5	8 x 1	0 x 7	8 x 6	5 x 8	9 x 5	4 x 2	1 x 5	5 x 5
G.	9 x 2	6 x 7	3 x 0	7 x 3	3 x 8	6 x 10	2 x 4	9 x 7	0 x 8	7 x 2
H.	7 x 0	1 x 4	8 x 7	5 x 7	6 x 0	9 x 1	3 x 9	8 x 8	6 x 8	10 x 3
I.	4 x 4	5 x 0	3 x 1	7 x 1	4 x 3	2 x 5	10 x 10	6 x 1	5 x 6	4 x 5
J.	8 x 10	2 x 6	8 x 0	6 x 9	9 x 0	10 x 7	9 x 3	1 x 10	3 x 2	8 x 9

Multiplication Sure-Shot

Be a sure-shot! Complete each multiplication chart below.

A)

X	2		8		3		0		6	
6	12	30						54		
5				5						
7						49				28

B)

X		8		6		4	
		64					24
5			35		25		
7	63						21
				54			

C)

X		5		8		9		6		7
8	72									
		30					30			
					63			54		
7			42						56	
						36				

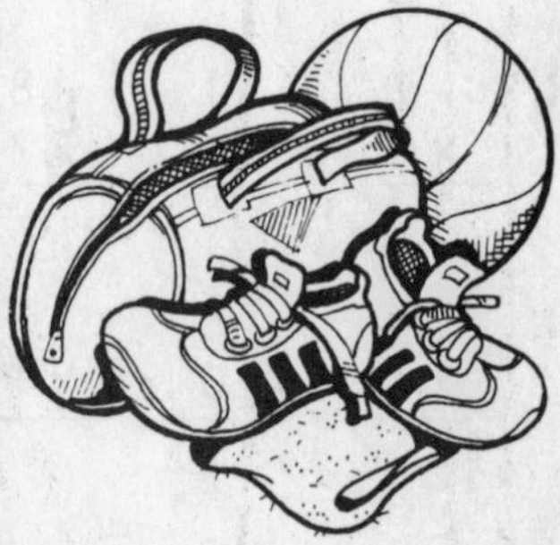

40	42	54	56	63
blue	orange	brown	red	yellow

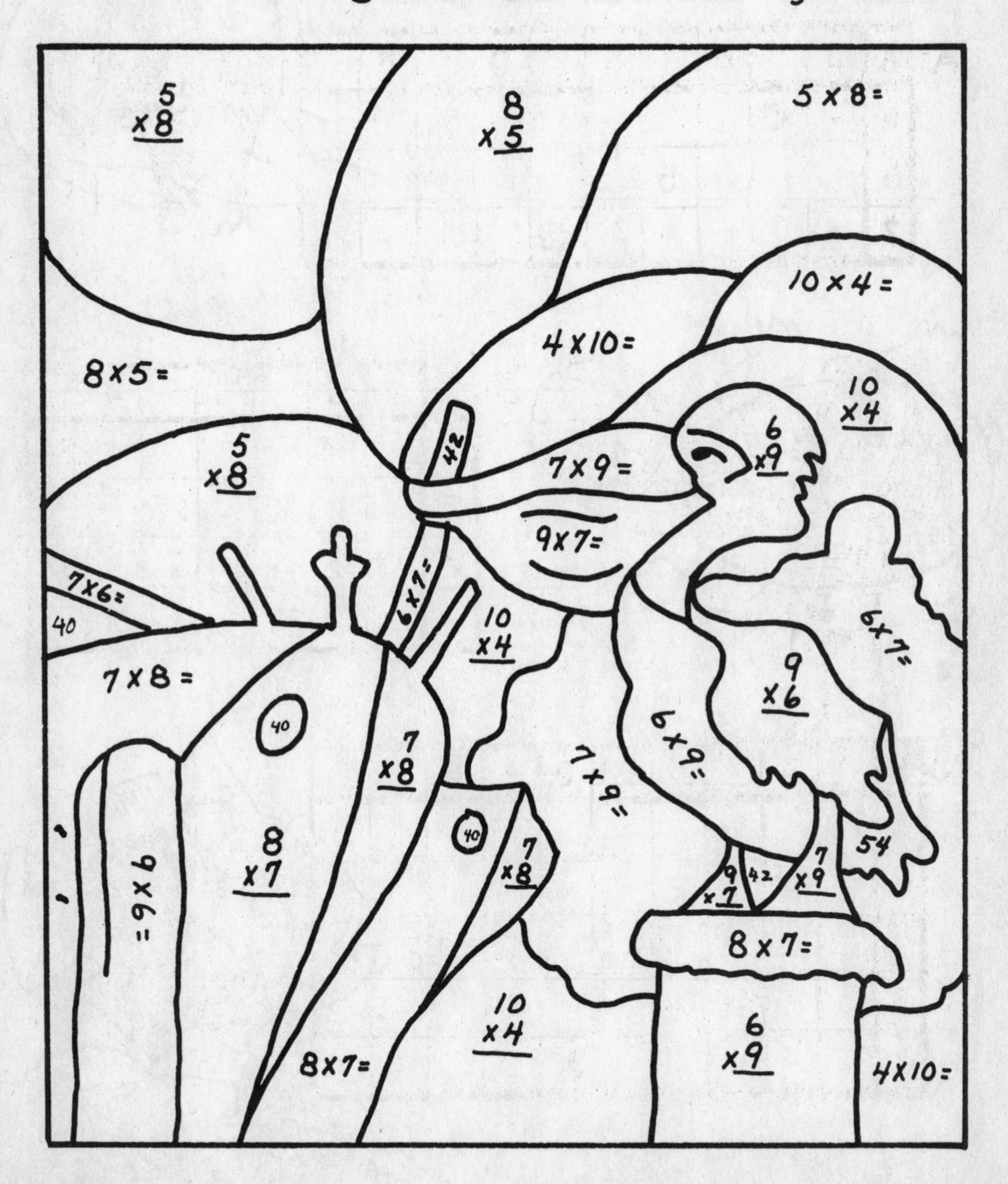

Sea Life

Multiply and write each product. Cross off the answer on the matching animal.

A.

10	10	10	10	10	10
x 3	x 7	x 9	x 4	x 10	x 5

10	10	10	10	10	10
x 11	x 1	x 8	x 2	x 6	x 12

A.

60	90	30
50	20	70
110	80	100
40	120	10

B.

11	11	11	11	11	11
x 7	x 3	x 9	x 10	x 2	x 6

11	11	11	11	11	11
x 11	x 4	x 12	x 1	x 8	x 5

C.

12	12	12	12	12	12
x 12	x 7	x 3	x 1	x 8	x 5

12	12	12	12	12	12
x 4	x 11	x 6	x 9	x 2	x 10

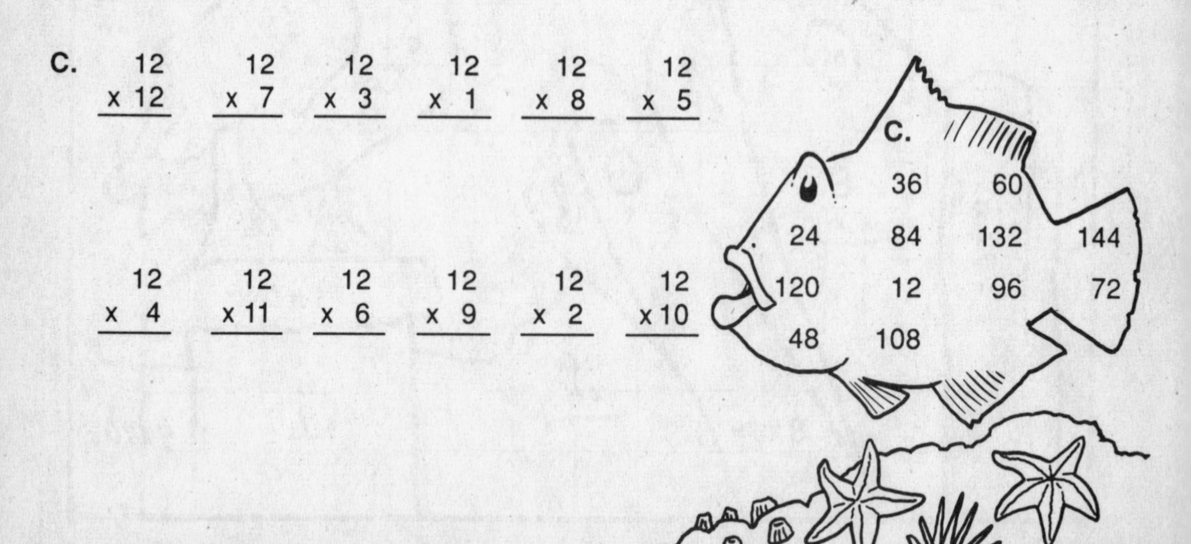

Fill in the missing numbers on the multiplication chart.

X	6	9		5	2	0		11			1		4
10									120				40
4			12								4		
	36				12								
7								77			7		
5		45								50			
								33					12
2	12		6									16	
12							84						
9					18				108				
11		99				0						88	
				0							0		
1							7		12				
8				40									32

A Tongue Twister:

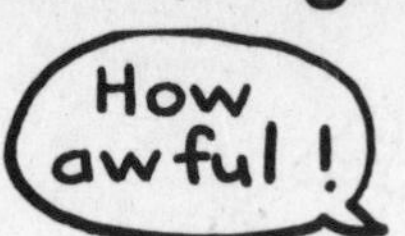

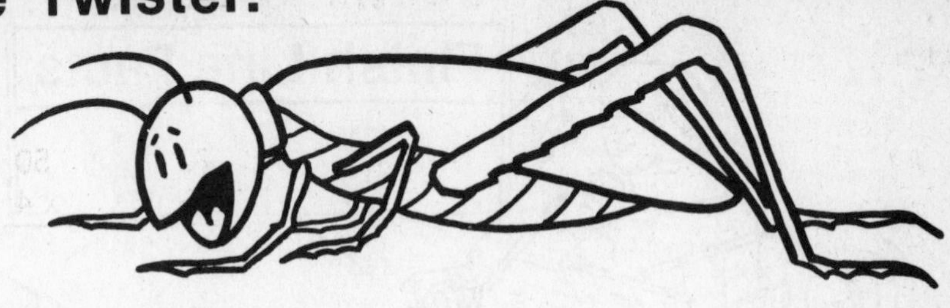

__ 55   __ 49 __ 48 __ 96 __ 96 __ 72 Y   __ 49 __ 81 __ 55 __ 0 __ 60

__ 49 __ 32 __ 27 __ 27 __ 33 __ 96 __ 72   U __ 15

__ 49 __ 55 __ 33 __ 33 __ 32 __ 0 __ 54   __ 32 __ 45   __ 49 __ 48 __ 96 __ 96 __ 0

__ 49 __ 48 __ 55 __ 54 __ 54 __ 36 __ 32 __ 15 __ 15 __ 96 __ 48 __ 54.

A 11×5	B 3×9	D 12×6	E 8×12	F 5×9
G 7×7	H 12×3	I 9×9	L 11×3	N 0×6
O 8×4	P 15×1	R 12×4	S 6×9	T 12×5

Pedaling Power

Multiply and write each product.

A. 60 x 8 = ______

B. 20 x 9 = ______

C. 7 x 70 = ______

D. 60 x 6 = ______

E. 90 x 6 = ______

F. 7 x 60 = ______

G. 50 x 6 = ______

H. 3 x 70 = ______

I. 70 x 6 = ______

J. 60 x 1 = ______

K. 20 x 3 = ______

L. 5 x 80 = ______

M. 60 x 2 = ______

N. 50 x 9 = ______

O. 90 x 2 = ______

P. 50 x 0 = ______

Q. 80 x 6 = ______

R. 30 x 4 = ______

S. 7 x 20 = ______

T. 30 x 8 = ______

U.	40 x 6	70 x 9	50 x 4	90 x 9	60 x 5
V.	40 x 7	60 x 4	80 x 9	70 x 1	20 x 2
W.	30 x 6	90 x 4	50 x 5	80 x 4	70 x 5
X.	40 x 8	80 x 7	70 x 4	50 x 3	90 x 3
Y.	30 x 3	20 x 4	20 x 5	60 x 9	70 x 8
Z.	30 x 9	80 x 2	70 x 2	80 x 8	90 x 7

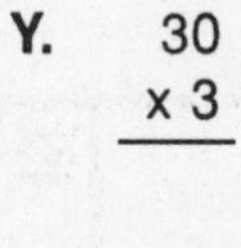

Brainwork! Double-check with mental math the multiplication you did above.

A Happy Thought:

___'___ ___ ___ ___ ___ ___ ___ ___

0 80 180 90 150 120 180 540 0

___ ___ ___'___ ___ ___ ___ ___

420 30 560 70 800 0 560 50

___ ___ ___ ___ ___ ___ ___ ___ ___ ___ ___ ___ ___.

70 0 80 60 70 90 800 60 60 360 180 30 50

A 10×3	B 50×3	C 70×6	D 10×5	E 20×3
F 200×4	I 50×0	L 90×4	M 40×2	N 80×7
O 30×3	S 20×9	T 70×1	U 12×10	Y 60×9

Can You Canoe?

Multiply and write each product.

A.	400 x 8	700 x 2	400 x 6	600 x 8	900 x 3
B.	800 x 7	400 x 5	900 x 8	700 x 6	700 x 9
C.	500 x 7	400 x 9	800 x 4	500 x 9	800 x 5
D.	900 x 2	500 x 3	600 x 9	900 x 5	700 x 3
E.	600 x 5	700 x 7	800 x 1	600 x 2	400 x 3
F.	900 x 9	400 x 7	200 x 4	800 x 2	600 x 6
G.	300 x 3	700 x 4	100 x 9	800 x 8	700 x 0

H. 200 x 2 = ________

I. 6 x 900 = ________

J. 6 x 800 = ________

K. 500 x 4 = ________

L. 3 x 200 = ________

M. 500 x 8 = ________

N. 4 x 400 = ________

O. 900 x 7 = ________

P. 600 x 4 = ________

Q. 2 x 500 = ________

R. 800 x 3 = ________

S. 800 x 9 = ________

T. 3 x 600 = ________

U. 5 x 700 = ________

V. 500 x 5 = ________

Brainwork! Double-check with mental math the multiplication you did above.

18	27	42, 49, 54	30 to 40
yellow	orange	brown	green

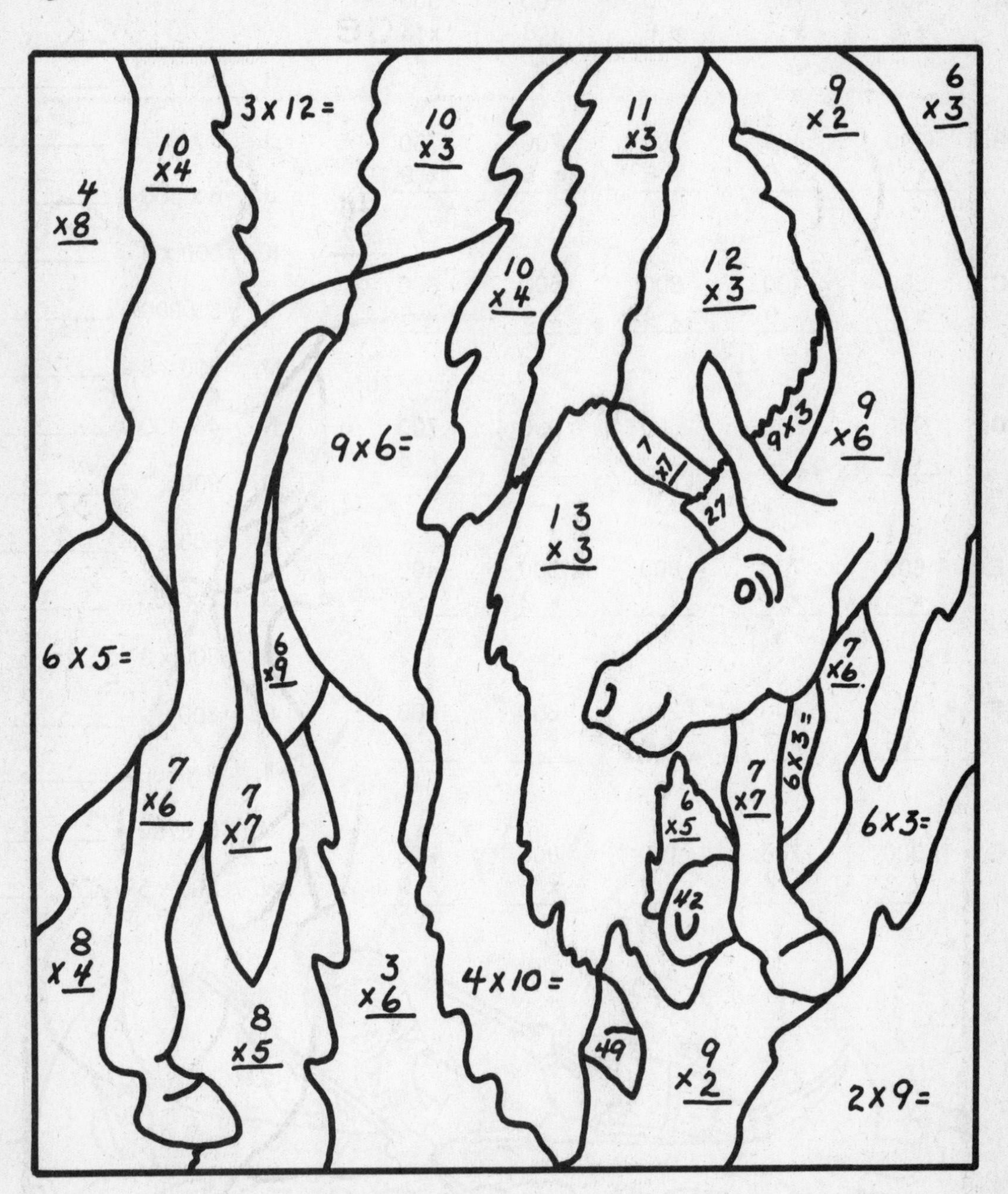

18 yellow | 24 red | 63, 64, 66 orange | 70 to 85 black

18	24	63, 64, 66	70 to 85
yellow	red	orange	black

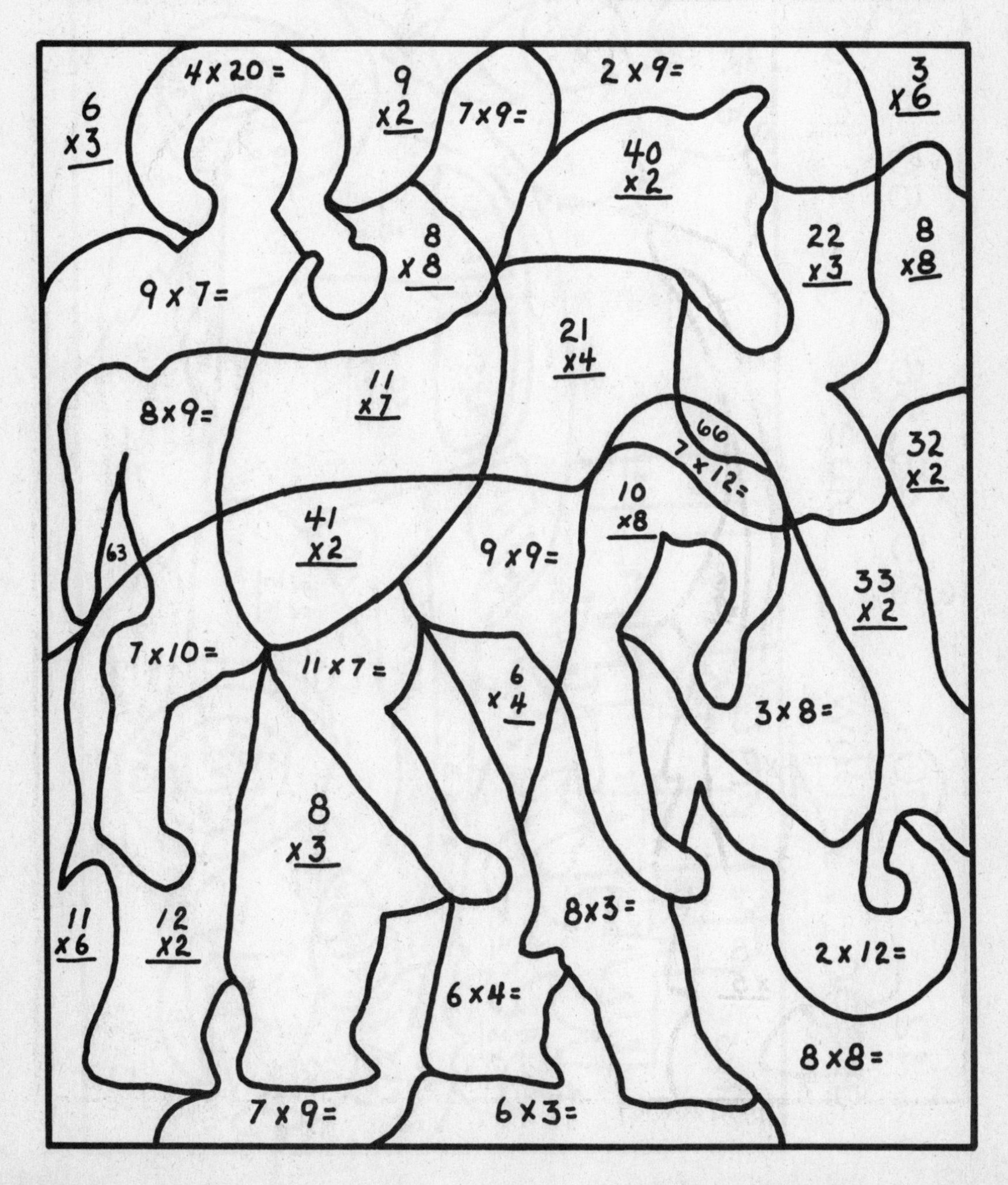

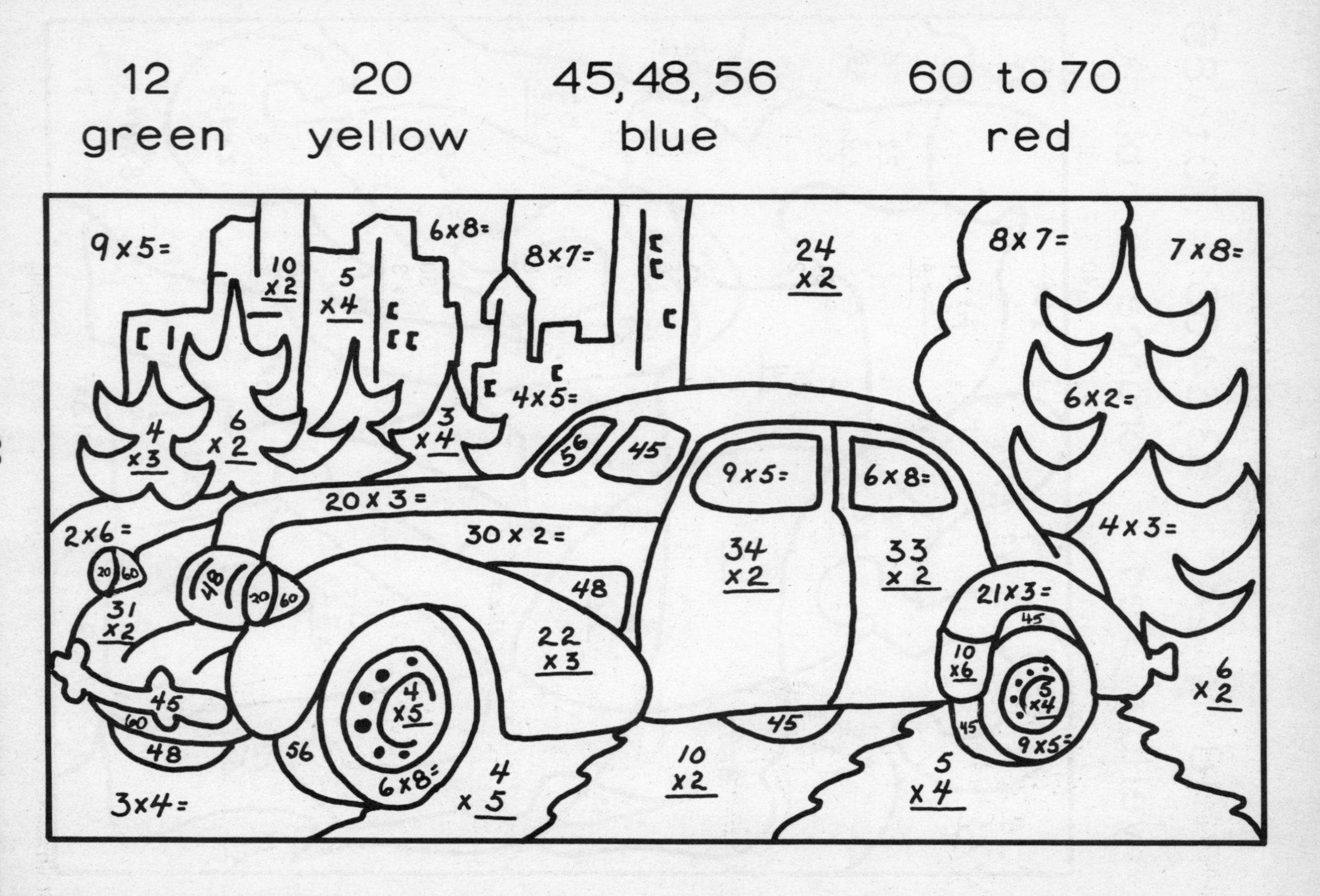
12 green
20 yellow
45, 48, 56 blue
60 to 70 red
9 x 5 =
10 x 2
5 x 4
6 x 8 =
8 x 7 =
24 x 2
8 x 7 =
7 x 8 =
4 x 3
6 x 2
3 x 4
4 x 5 =
6 x 2 =
56
45
9 x 5 =
6 x 8 =
20 x 3 =
2 x 6 =
30 x 2 =
4 x 3 =
20
60
48
48
34 x 2
33 x 2
31 x 2
21 x 3 =
45
22 x 3
10 x 6
45
6 x 2
60
4 x 5
5 x 4
45
45
48
56
9 x 5 =
10 x 2
6 x 8 =
4 x 5
5 x 4
3 x 4 =

Batter Up!

Multiply and write each product.

A.	32 x 2	51 x 3	40 x 9	21 x 7	30 x 8
B.	21 x 2	92 x 4	21 x 4	31 x 7	43 x 2
C.	51 x 4	82 x 2	51 x 2	92 x 3	60 x 7
D.	81 x 5	30 x 5	41 x 6	20 x 9	71 x 8
E.	70 x 4	43 x 2	91 x 2	21 x 6	91 x 8
F.	40 x 4	61 x 5	30 x 3	33 x 3	23 x 3
G.	60 x 6	21 x 9	40 x 2	81 x 8	90 x 7

Brainwork! Write a multiplication word problem about baseball.

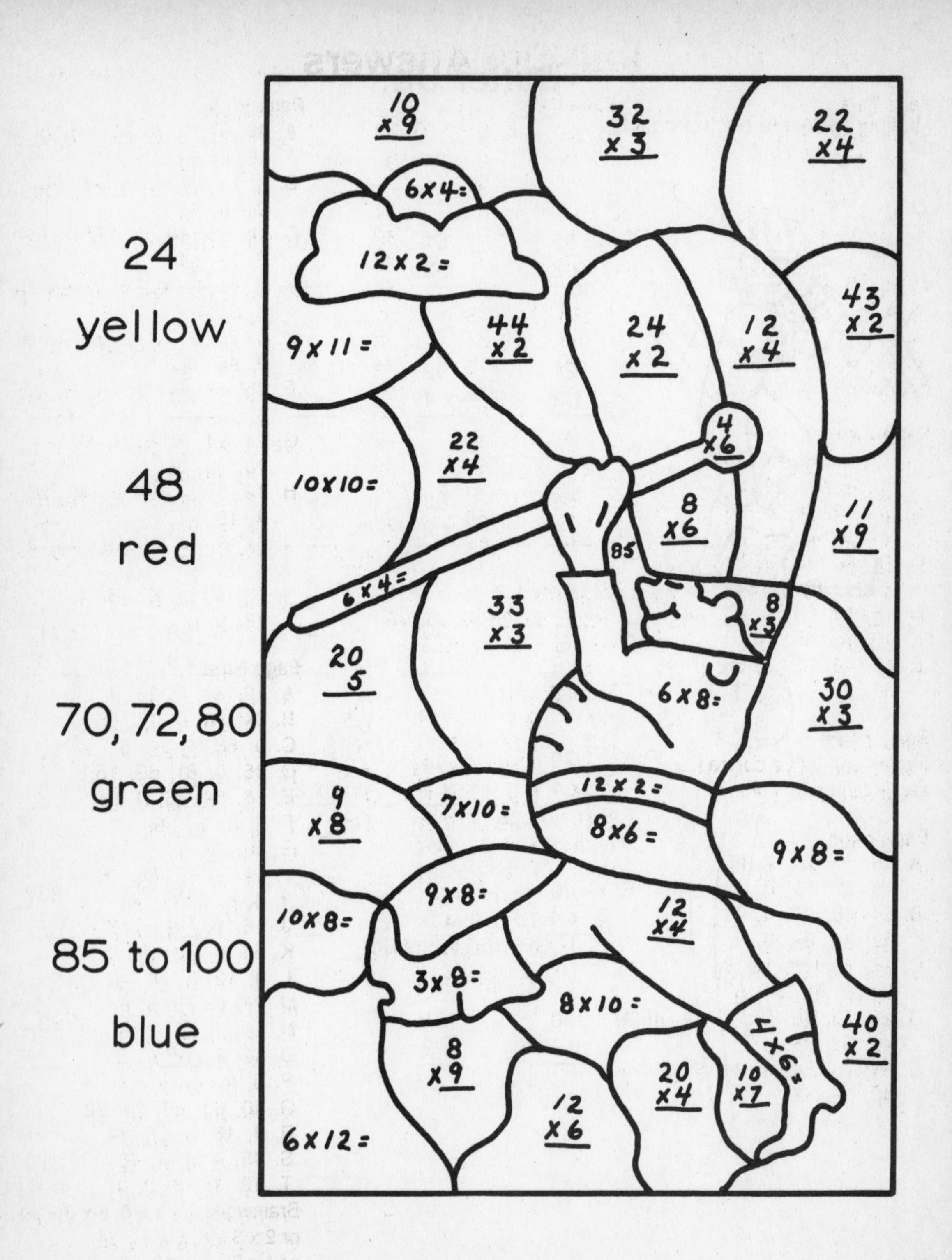
24
yellow
48
red
70, 72, 80
green
85 to 100
blue
10 × 9
32 × 3
22 × 4
6 × 4 =
12 × 2 =
43 × 2
9 × 11 =
44 × 2
24 × 2
12 × 4
4 × 6
22 × 4
10 × 10 =
8 × 6
11 × 9
85
6 × 4 =
33 × 3
8 × 3
20 5
6 × 8 =
30 × 3
12 × 2 =
9 × 8
7 × 10 =
8 × 6 =
9 × 8 =
9 × 8 =
10 × 8 =
12 × 4
3 × 8 =
8 × 10 =
4 × 6 =
40 × 2
8 × 9
20 × 4
10 × 7
12 × 6
6 × 12 =

Pull-Out Answers

Page One

A. 0, 0, 0, 0, 0
B. 0, 0, 0, 0, 0
C. 9, 4, 0, 3, 7
D. 2, 1, 8, 6, 5
E. 16, 6, 10, 0, 2
F. 12, 18, 14, 8, 4
G. 0, 15, 21, 3, 9
H. 6, 27, 18, 12, 24
I. 21, 16, 0, 10, 18
J. 27, 6, 2, 12, 14

Brainwork! 8 games

Page Two

A. 32, 16, 8, 12, 28
20, 4, 24, 36, 0
B. 35, 15, 25, 5, 10
30, 20, 0, 45, 40
C. 18, 30, 12, 36, 42
6, 24, 48, 0, 54
D. 12, 12, 54, 4, 40, 28,
42, 15, 35, 16
E. 10, 48, 36, 5, 36, 0, 18,
32, 25, 30

Page Three

Picture should be colored according to the code.

Page Four

A. 49, 21, 28, 7, 56
35, 63, 14, 0, 42
B. 24, 56, 40, 8, 72
32, 48, 64, 0, 16
C. 45, 63, 27, 18, 36
9, 72, 81, 54, 0
D. 21, 18, 48, 56, 0, 42, 45
49, 32, 27
E. 64, 28, 54, 8, 63, 81, 14
40, 36, 72

Page Five

A.	B.	C.
64	6	20
25	8	0
72	54	0
16	0	81
9	0	40
12	7	36
5	63	45
54	45	21
14	24	15
30	18	24
35	32	28
18	0	35
56	10	42
42	36	48
49	30	40
48	56	27
72	63	

Page Six

Picture should be colored according to the code.

Page Seven

A-4	H-6	O-40	V-0
B-9	I-32	P-7	W-63
C-12	J-25	Q-24	X-49
D-64	K-9	R-20	Y-16
E-15	L-35	S-30	Z-45
F-28	M-8	T-48	
G-36	N-54	U-18	

1. A LIGHTHOUSE
2. ALL OF THEM
3. TOGETHER WE MAKE MORE CENTS.

Page Eight

A. 48, 27, 2, 0, 32, 0, 35,
18, 0, 63
B. 8, 1, 54, 30, 0, 42, 28,
24, 0, 45
C. 14, 35, 36, 0, 12, 3, 16,
6, 18, 54
D. 4, 25, 8, 42, 5, 7, 40, 0,
0, 21
E. 0, 28, 9, 36, 2, 8, 10, 0,
9, 48
F. 45, 72, 27, 9, 56, 16, 6,
8, 30, 0
G. 24, 24, 3, 20, 6, 0, 21,
18, 40, 36
H. 12, 0, 14, 6, 32, 49, 25,
4, 16, 81
I. 64, 0, 18, 10, 56, 16, 20,
4, 7, 72
J. 63, 5, 24, 0, 15, 4, 12,
12, 0, 0

Page Nine

A. 12, 30, 0, 20, 7
B. 64, 20, 0, 18, 56
C. 0, 56, 8, 24, 0
D. 45, 0, 81, 63, 15
E. 16, 54, 36, 0, 24
F. 6, 0, 15, 48, 1
G. 25, 18, 7, 9, 35
H. 14, 0, 12, 36, 10
I. 0, 8, 2, 14, 48
J. 16, 18, 21, 5, 3
K. 54, 0, 28, 27, 0
L. 4, 12, 5, 30, 49
M. 27, 8, 72, 8, 6
N. 32, 42, 0, 2, 6
O. 42, 6, 24, 0, 36
P. 3, 0, 40, 21, 9
Q. 40, 63, 35, 10, 72
R. 4, 16, 0, 28, 24
S. 45, 0, 9, 4, 32
T. 12, 18, 0, 0, 0

Brainwork! 4 x 2 = 8, 8 x 3 = 24
or 2 x 3 = 6, 6 x 4 = 24
or 4 x 3 = 12, 12 x 2 = 24

Pull-Out Answers

Page Ten

x	2	5	8	6	4	0	3	1	7	9
2	4	10	16	12	8	0	6	2	14	18
3	6	15	24	18	12	0	9	3	21	27
4	8	20	32	24	16	0	12	4	28	36
8	16	40	64	48	32	0	24	8	56	72

x	6	8	1
4	24	32	4
7	42	56	7
5	30	40	5
6	36	48	6
3	18	24	3
2	12	16	2
9	54	72	9
8	48	64	8
0	0	0	0
1	6	8	1

x	5	4	7	3	2	8	9	6	0	1
4	20	16	28	12	8	32	36	24	0	4
7	35	28	49	21	14	56	63	42	0	7
9	45	36	63	27	18	72	81	54	0	9

Page Eleven

A. 10, 9, 15, 24, 0, 90, 42, 32, 20, 8
B. 81, 18, 32, 7, 49, 15, 2, 45, 28, 18
C. 16, 54, 12, 16, 0, 30, 63, 20, 6, 35
D. 18, 36, 0, 5, 21, 18, 0, 0, 40, 24
E. 28, 9, 56, 4, 36, 24, 40, 12, 72, 6
F. 14, 50, 8, 0, 48, 40, 45, 8, 5, 25
G. 18, 42, 0, 21, 24, 60, 8, 63, 0, 14
H. 0, 4, 56, 35, 0, 9, 27, 64, 48, 30
I. 16, 0, 3, 7, 12, 10, 100, 6, 30, 20
J. 80, 12, 0, 54, 0, 70, 27, 10, 6, 72

Page Twelve

A)

X	2	5	8	1	3	7	0	9	6	4
6	12	30	48	6	18	42	0	54	36	24
5	10	25	40	5	15	35	0	45	30	20
7	14	35	56	7	21	49	0	63	42	28

B)

X	9	8	7	6	5	4	3
8	72	64	56	48	40	32	24
5	45	40	35	30	25	20	15
7	63	56	49	42	35	28	21
9	81	72	63	54	45	36	27

C)

X	9	5	6	8	7	9	5	6	8	7
8	72	40	48	64	56	72	40	48	64	56
6	54	30	36	48	42	54	30	36	48	42
9	81	45	54	72	63	81	45	54	72	63
7	63	35	42	56	49	63	35	42	56	49
4	36	20	24	32	28	36	20	24	32	28

Page Thirteen

Picture should be colored according to the code.

Page Fourteen

A. 30, 70, 90, 40, 100, 50 110, 10, 80, 20, 60, 120
B. 77, 33, 99, 110, 22, 66 121, 44, 132, 11, 88, 55
C. 144, 84, 36, 12, 96, 60 48, 132, 72, 108, 24, 120

Page Fifteen

X	6	9	3	5	2	0	7	11	12	10	1	8	4
10	60	90	30	50	20	0	70	110	120	100	10	80	40
4	24	36	12	20	8	0	28	44	48	40	4	32	16
6	36	54	18	30	12	0	42	66	72	60	6	48	24
7	42	63	21	35	14	0	49	77	84	70	7	56	28
5	30	45	15	25	10	0	35	55	60	50	5	40	20
3	18	27	9	15	6	0	21	33	36	30	3	24	12
2	12	18	6	10	4	0	14	22	24	20	2	16	8
12	72	108	36	60	24	0	84	132	144	120	12	96	48
9	54	81	27	45	18	0	63	99	108	90	9	72	36
11	66	99	33	55	22	0	77	121	132	110	11	88	44
0	0	0	0	0	0	0	0	0	0	0	0	0	0
1	6	9	3	5	2	0	7	11	12	10	1	8	4
8	48	72	24	40	16	0	56	88	96	80	8	64	32

Page Sixteen

A GREEDY GIANT GOBBLED UP GALLONS OF GREEN GRASSHOPPERS.

Page Seventeen

A. 480
B. 180
C. 490
D. 360
E. 540
F 420
G. 300
H. 210
I. 420
J. 60
K. 60
L. 400
M. 120
N. 450
O. 180
P. 0
Q. 480
R. 120
S. 140
T. 240
U. 240, 630, 200, 810, 300
V. 280, 240, 720, 70, 40
W. 180, 360, 250, 320, 350
X. 320, 560, 280, 150, 270
Y. 90, 80, 100, 540, 560
Z. 270, 160, 140, 640, 630

Page Eighteen

I'M SO BUSY I CAN'T FIND TIME TO FEEL SAD.

Page Nineteen

A. 3,200; 1,400; 2,400; 4,800; 2,700
B. 5,600; 2,000; 7,200; 4,200; 6,300
C. 3,500; 3,600; 3,200; 4,500; 4,000
D. 1,800; 1,500; 5,400; 4,500; 2,100
E. 3,000; 4,900; 800; 1,200; 1,200
F. 8,100; 2,800; 800; 1,600; 3,600
G. 900; 2,800; 900; 6,400; 0
H. 400
I. 5,400
J. 4,800
K. 2,000
L. 600
M. 4,000
N. 1,600
O. 6,300
P. 2,400
Q. 1,000
R. 2,400
S. 7,200
T. 1,800
U. 3,500
V. 2,500

Page Twenty

Picture should be colored according to the code.

Pull-Out Answers

Page Twenty-one
Picture should be colored according to the code.

Page Twenty-two
Picture should be colored according to the code.

Page Twenty-three
A. 64, 153, 360, 147, 240
B. 42, 368, 84, 217, 86
C. 204, 164, 102, 276, 420
D. 405, 150, 246, 180, 568
E. 280, 86, 182, 126, 728
F. 160, 305, 90, 99, 69
G. 360, 189, 80, 648, 630

Page Twenty-four
Picture should be colored according to the code.

Page Twenty -five
ALL THE WONDERS YOU SEEK ARE HIDDEN WITHIN.

Page Twenty-six
WHEN IT BUMPED INTO AN ELECTRIC EEL.

Page Twenty-seven
A. 282, 272, 330, 413
B. 225, 534, 174, 268
C. 306, 320, 208, 440, 603, 504
D. 258, 371, 110, 801, 315, 504
E. 576, 784, 238, 84, 136, 225
F. 65, 438, 441, 469
G. 52, 138, 92, 72
The fourth problem in Row D should be colored red.

Page Twenty-eight
CAN YOU KICK SIX SHORT STICKS QUICKLY?

Page Twenty-nine
A. 152; 4 x 40 = 160
B. 87; 30 x 3 = 90
C. 306; 6 x 50 = 300
D. 328; 8 x 40 = 320
E. 639; 70 x 9 = 630
F. 119; 7 x 20 = 140
G. 135; 30 x 5 =150
H. 86; 2 x 40 = 80
I. 80; 5 x 20 = 100
J. 406; 60 x 7 = 420
K. 148; 4 x 40 = 160
L. 78; 3 x 30 = 90
M. 504; 60 x 8 = 480
N. 154; 20 x 7 = 140
O. 376; 50 x 8 = 400

Page Thirty
BOWLING. YOU CAN HEAR A PIN DROP!

Page Thirty-one
YOUR HAPPY SMILE MAKES ME FEEL GLAD.

Page Thirty-two
A. 5,052; 2,490
B. 1,268; 1,101; 6,584
C. 5,736; 4,872; 2,046; 5,684
D. 1,593; 1,396; 5,064; 3,750; 5,265; 1,888
E. 1,587; 4,480; 3,575; 3,951; 1,724; 2,232
F. 5,019; 2,185; 2,478; 6,664; 4,192; 6,244
G. 3,020; 5,593; 3,549; 8,433; 5,478; 5,208
Brainwork!
850 + 850 + 850 = 2,550

Page Thirty-three
THE MIGHTY MISSISSIPPI RIVER, OF COURSE!

Page Thirty-four
BECAUSE THE BATS CAN'T COME OUT UNTIL DARK.

Page Thirty-five
SHE SHEARED SIX SHEEP TO MAKE SEVEN SHIRTS.

Page Thirty-six
A SUPER MAGNETIC POINTER-PICKLE!

Page Thirty-seven
FREEDOM IS THE KEY TO OUR FRIENDSHIP.

Page Thirty-eight
MY WORK IS SO SECRET EVEN I DON'T KNOW WHAT I'M DOING!

Page Thirty-nine
A. $32.00, $3.36, $5.65, $35.55, $15.54, $38.25
B. $9.12, $68.31, $67.50, $3.75, $19.35, $63.42
C. $57.26, $58.50, $1.78, $111.87, $20.00, $37.20
D. $80.00, $64.72, $3.78, $4.70, $60.68, $52.80
E. $3.84, $14.90, $207.00, $.81, $329.22, $26.25
Brainwork! $754.65

Pull-Out Answers

Page Forty

X	4	1	10	8	0	9	6	2	12	7	3	11	5
7	28	7	70	56	0	63	42	14	84	49	21	77	35
5	20	5	50	40	0	45	30	10	60	35	15	55	25
2	8	2	20	16	0	18	12	4	24	14	6	22	10
1	4	1	10	8	0	9	6	2	12	7	3	11	5
10	40	10	100	80	0	90	60	20	120	70	30	110	50
0	0	0	0	0	0	0	0	0	0	0	0	0	0
8	32	8	80	64	0	72	48	16	96	56	24	88	40
11	44	11	110	88	0	99	66	22	132	77	33	121	55
3	12	3	30	24	0	27	18	6	36	21	9	33	15
9	36	9	90	72	0	81	54	18	108	63	27	99	45
6	24	6	60	48	0	54	36	12	72	42	18	66	30
12	48	12	120	96	0	108	72	24	144	84	36	132	60
4	16	4	40	32	0	36	24	8	48	28	12	44	20

Page Forty-one
THREE HUNGRY FROGS ON A SUMMER PICNIC.

Page Forty-two
THE WIND WHIPS WAVES OVER THE WET WHARF.

Page Forty-three
OPEN THE DOOR WITH A SKELETON KEY.

Page Forty-four
A. 4,851; 1,435; 1,540; 2,136; 2,162; 3,712
B. 4,425; 1,290; 9,310; 6,916; 2,160; 5,475
C. 2,124; 1,608; 2,208; 4,466; 6,958; 7,905

Page Forty-five
A–C. 713; 5,220; 1,060
D–G. 3,266; 2,288; 5,248; 4,760
H–K. 3,948; 1,825; 3,894; 450
L–O. 3,936; 5,070; 1,404; 9,801

Brainwork! 12 x 43 = 516
21 x 34 = 714 21 x 34 > 12 x 43

Page Forty-six
I LIKE HAVING YOU FOR A FRIEND.

Page Forty-seven
1. 5
2. 10
3. 9
4. 3
5. 6
6. 4
7. 9
8. 10
9. 8
10. 5
11. 8
12. 10
13. 4, 2, 2, 2
14. 9, 2, 3, 3
15. 6, 3, 4, 2
16. 8, 8, 2, 4
17. 4, 2, 2, 2
18. 6, 2, 3, 2
19. 4, 4, 2, 2
20. 9, 2, 3, 3
21. 8, 2, 4, 3
22. 9, 9, 3, 3
23. 4, 10, 5, 2
24. 6, 4, 2, 2 or 3, 8, 1, 4
25. 8, 2, 2, 4

Page Forty-eight
A–E. 17,514; 5,236; 31,008; 13,266; 72,990
F–J. 12,443; 42,328; 15,939; 18,647; 41,470
K–O. 29,681; 4,144; 78,540; 21,632; 15,288
P–T. 24,055; 23,959; 45,568; 32,652; 19,200

A Happy Thought:

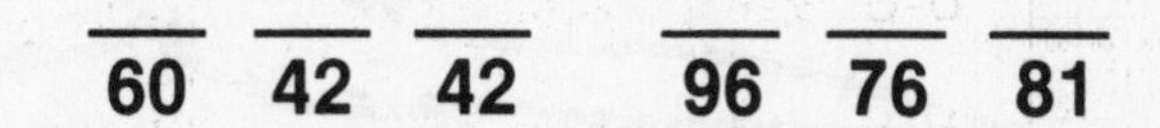

60 42 42 96 76 81

0 49 75 72 81 45 65 70 49 51

65 81 81 52 60 45 81

76 84 72 72 81 75 0 84 96 76 84 75.

A	D	E	H	I
15 × 4	12 × 6	27 × 3	38 × 2	12 × 7

K	L	N	O	R
26 × 2	14 × 3	25 × 3	49 × 1	15 × 3

S	T	U	W	Y
13 × 5	12 × 8	17 × 3	13 × 0	14 × 5

Riddle: When was the octopus shocked?

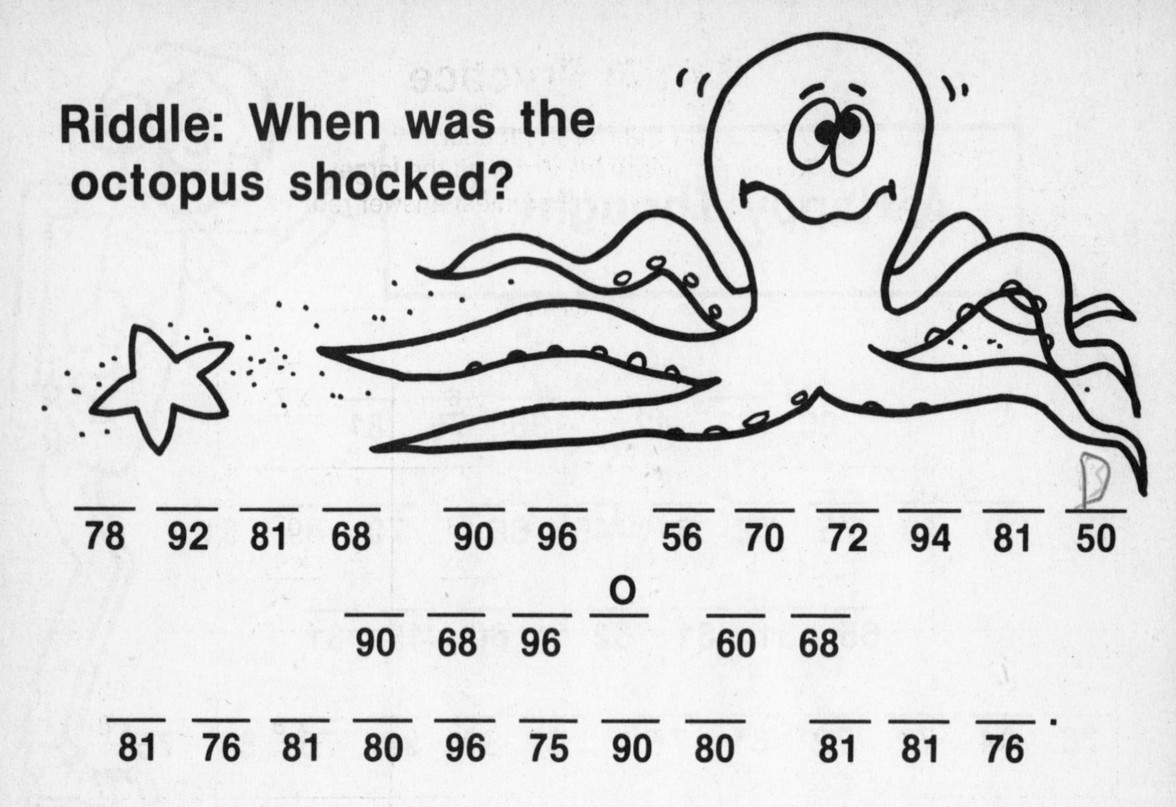

___ ___ ___ ___   ___ ___   ___ ___ ___ ___ ___ D
78 92 81 68   90 96   56 70 72 94 81 50

___ ___ ___ O   ___ ___
90 68 96   60 68

___ ___ ___ ___ ___ ___ ___ ___   ___ ___ ___.
81 76 81 80 96 75 90 80   81 81 76

A 12×5 = 60	B 14×4 = 56	C 16×5 = 80	D 25×2 = 50	E 27×3 = 81
H 23×4 = 92	I 15×6 = 90	L 19×4 = 76	M 12×6 = 72	N 17×4 = 68
P 47×2 = 94	R 15×5 = 75	T 12×8 = 96	U 35×2 = 70	W 26×3 = 78

Target Practice

Multiply and write each product.
To find out where the arrow hit the target,
color the box with the greatest answer red.

	A. 47 x 6	34 x 8	55 x 6	59 x 7	
	B. 45 x 5	89 x 6	58 x 3	67 x 4	
C. 34 x 9	64 x 5	26 x 8	55 x 8	67 x 9	63 x 8
D. 43 x 6	53 x 7	22 x 5	89 x 9	35 x 9	72 x 7
E. 72 x 8	98 x 8	34 x 7	12 x 7	17 x 8	25 x 9
	F. 13 x 5	73 x 6	49 x 9	67 x 7	
	G. 13 x 4	69 x 2	23 x 4	12 x 6	

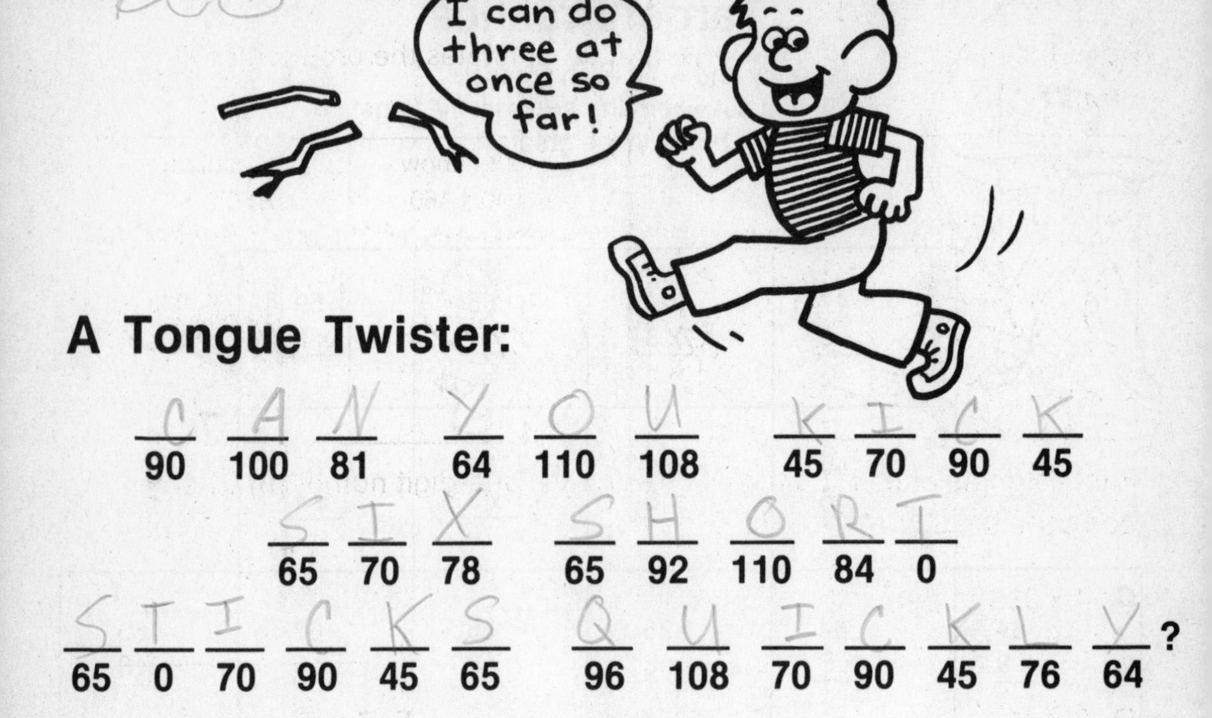

A Tongue Twister:

___ ___ ___ ___ ___ ___ ___ ___ ___ ___
90 100 81 64 110 108 45 70 90 45

___ ___ ___ ___ ___ ___ ___ ___
65 70 78 65 92 110 84 0

___ ___ ___ ___ ___ ___ ___ ___ ___ ___ ___ ___ ___ ?
65 0 70 90 45 65 96 108 70 90 45 76 64

A 25×4	C 15×6	H 23×4	I 14×5	K 15×3
L 38×2	N 27×3	O 22×5	Q 12×8	R 28×3
S 13×5	T 43×0	U 12×9	X 39×2	Y 16×4

Estimating With Mouse

Read the cartoon to find out how Mouse estimates the product for Problem A.

First estimate each product. (Do not round one-digit numbers.) Then compute the actual product.

A. 4 x 38 = 152
4 X 40 = 160

B. 29 x 3 = ______

C. 6 x 51 = ______

D. 8 x 41 = ______

E. 71 x 9 = ______

F. 7 x 17 = ______

G. 27 x 5 = ______

H. 2 x 43 = ______

I. 5 x 16 = ______

J. 58 x 7 = ______

K. 4 x 37 = ______

L. 3 x 26 = ______

M. 63 x 8 — 60 X 8 = 480

N. 22 x 7

O. 47 x 8

Try This! Draw a cartoon with four frames to show how you decided on the estimate for problem N or O.

Riddle: What is the quietest game in the world?

		W				.			
969	999		1896	846	1688	0	1209	999	1280

848	468	1688	1284	215	468	880	468

						!
800	846	1688	1866	880	999	800

A 234×2	B 323×3	C 212×4	D 622×3	E 215×1
G 525×0	H 321×4	I 423×2	L 632×3	N 422×4
O 111×9	P 200×4	R 440×2	U 320×4	Y 403×3

A Happy Thought:

___ O ___ ___ ___ ___ ___ ___ ___
1236   1408 906   1269 488 693 693 1236

___ ___ ___ ___ ___ ___ ___ ___ ___ ___
1569 0 996 960 309   0 488 248 309 1569

___ ___ ___ ___ ___ ___
0 309   286 309 309 960

___ ___ ___ ___!
700 960 488 1284

A 244 × 2	D 321 × 4	E 103 × 3	F 143 × 2	G 100 × 7
H 423 × 3	I 332 × 3	K 124 × 2	L 320 × 3	M 506 × 0
P 231 × 3	R 302 × 3	S 523 × 3	U 704 × 2	Y 412 × 3

Multiplication Mountaineering

Multiply and write each product.

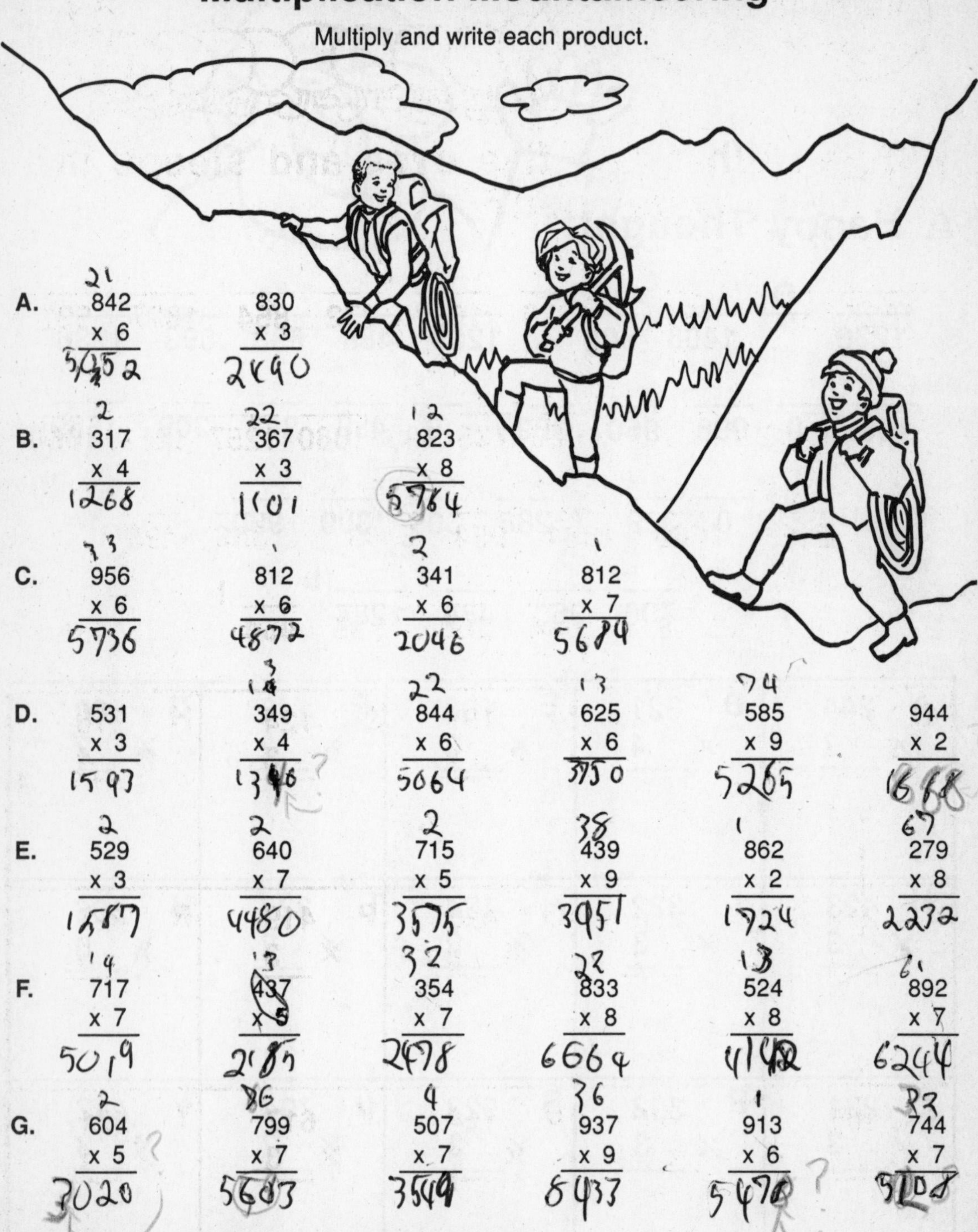

A.	842 x 6	830 x 3				
B.	317 x 4	367 x 3	823 x 8			
C.	956 x 6	812 x 6	341 x 6	812 x 7		
D.	531 x 3	349 x 4	844 x 6	625 x 6	585 x 9	944 x 2
E.	529 x 3	640 x 7	715 x 5	439 x 9	862 x 2	279 x 8
F.	717 x 7	437 x 5	354 x 7	833 x 8	524 x 8	892 x 7
G.	604 x 5	799 x 7	507 x 7	937 x 9	913 x 6	744 x 7

Brainwork! Write 3 x 850 as an addition problem.

Riddle : What has five eyes and sleeps in a water bed?

T H E — 1827 864 954

M I G H T Y — 1278 1060 372 864 1827 521

M I S S I S S I P P I — 1278 1060 735 735 1060 735 735 1060 1257 1257 1060

R I V E R , O F — 2085 1060 1851 954 2085 , 852 428

C O U R S E ! — 645 852 684 2085 735 954 !

C 215 × 3 = 645	E 318 × 3 = 954	F 107 × 4 = 428	G 124 × 3 = 372	H 216 × 4 = 864
I 212 × 5 = 1060	M 213 × 6 = 1278	O 426 × 2 = 852	P 419 × 3 = 1257	R 417 × 5 = 2085
S 105 × 7 = 735	T 203 × 9 = 1827	U 114 × 6 = 684	V 617 × 3 = 1851	Y 521 × 1 = 521

Riddle: Why can't ghosts play baseball on Saturday afternoons?

___ ___ ___ ___ ___ ___ ___
1284 1545 3690 642 2754 0 1545

___ ___ ___ ___ ___ ___ ___
2052 2178 1545 ' 1284 642 2052 0

___ ___ ___ ___ ___ ___ ___ ___
3690 642 3060 2052 3690 456 2884 1545

___ ___ ___ ___ ___ ___ ___ ___
456 2754 2052 2754 3060 2052 2076 2475

___ ___ ___ K .
1092 642 540

A 214 × 3	B 428 × 3	C 615 × 6	D 546 × 2	E 309 × 5
H 726 × 3	I 519 × 4	L 825 × 3	M 412 × 7	N 612 × 5
O 456 × 1	R 108 × 5	S 397 × 0	T 513 × 4	U 918 × 3

A Tongue Twister:

___ ___ ___   ___ ___ ___ ___ ___ ___ ___
1025 1227 1074   1025 1227 1074 1884 4065 1074 1585

___ ___ ___   ___ ___ ___ ___ ___
1025 3591 0   1025 1227 1074 1074 3078

___ ___   ___ ___ ___ ___
2075 2585   536 1884 2476 1074

___ ___ ___ ___ ___
1025 1074 2412 1074 1595

___ ___ ___ ___ ___ ___ .
1025 1227 3591 4065 2075 1025

A 314×6	D 317×5	E 537×2	H 409×3	I 513×7
K 619×4	M 536×1	N 319×5	O 517×5	P 513×6
R 813×5	S 205×5	T 415×5	V 402×6	X 759×0

Riddle: What is green and always points north?

___ 651   ___ 1074 ___ 2853 ___ 2540 ___ 1570 ___ 988

___ 3204 ___ 651 ___ 1136 ___ 935 ___ 1570 ___ 0 ___ 3615 ___ 1436 —

___ 2540 ___ 714 ___ 3615 ___ 935 ___ 0 ___ 1570 ___ 988

___ 2540 ___ 3615 ___ 1436 ___ 5778 ___ 756 ___ 1570 !

A 217×3	C 359×4	E 314×5	G 142×8	I 723×5
K 642×9	L 126×6	M 534×6	N 935×1	O 238×3
P 635×4	R 247×4	S 358×3	T 235×0	U 317×9

A Happy Thought:

___ ___ ___ ___ ___ ___ ___   ___ ___
374 889 468 468 936 718 1561   584 747

___ ___ ___   ___ ___ ___ ___ ___
1074 612 468   2285 468 456 1074 718

___ ___ ___
718 725 889

___ ___ ___ ___ ___ ___ ___ ___ ___ ___.
374 889 584 468 1056 936 747 612 584 708

D 234 × 4	E 156 × 3	F 187 × 2	H 153 × 4	I 146 × 4
K 457 × 5	M 223 × 7	N 132 × 8	O 359 × 2	P 236 × 3
R 127 × 7	S 249 × 3	T 358 × 3	U 145 × 5	Y 456 × 1

A Happy Thought:

___ ___   ___ ___ ___ ___   ___ ___
2325 0   756 777 2024 4473   1216 942

___ ___   ___ ___ ___ ___ ___ ___
942 777   942 2136 918 2024 2136 1128

___ V ___ ___   ___   D ___ ___ ' ___
2136   2136 936   1216   777 936 1128

___ ___ ___ ___   ___ ___ ___ ___
4473 936 777 756   756 1436 615 1128

___ ' ___   D ___ ___ ___ ___ !
1216 2325   777 1216 936 1645

A 123×5	C 459×2	E 356×6	G 235×7	H 359×4
I 152×8	K 639×7	M 465×5	N 234×4	O 259×3
R 506×4	S 157×6	T 376×3	W 378×2	Y 630×0

Suiting Up

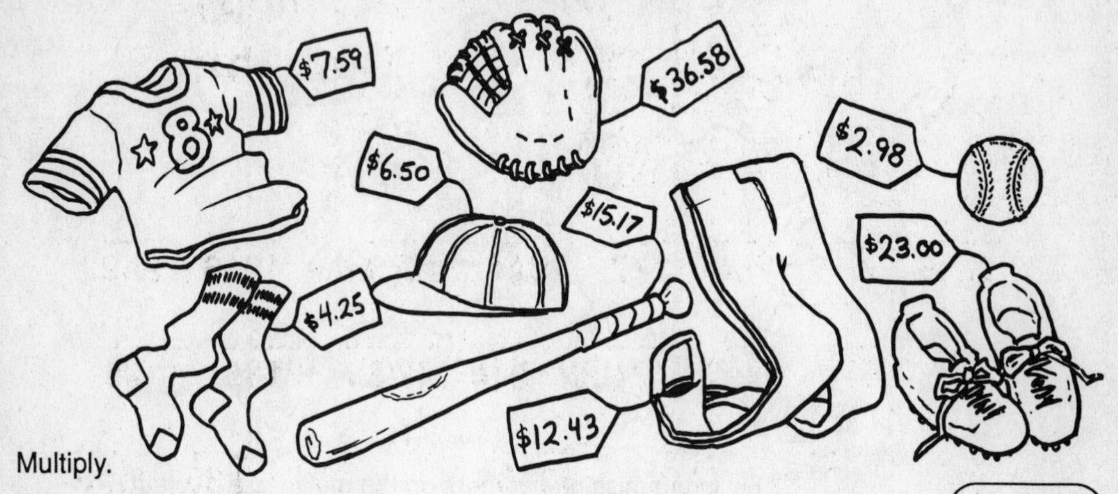

Multiply.

A.	$8.00 x 4	$.56 x 6	$1.13 x 5	$3.95 x 9	$7.77 x 2	$4.25 x 9
B.	$1.52 x 6	$7.59 x 9	$11.25 x 6	$.75 x 5	$6.45 x 3	$9.06 x 7
C.	$8.18 x 7	$6.50 x 9	$.89 x 2	$12.43 x 9	$5.00 x 4	$4.65 x 8
D.	$10.00 x 8	$8.09 x 8	$.63 x 6	$.94 x 5	$15.17 x 4	$26.40 x 2
E.	$1.28 x 3	$2.98 x 5	$23.00 x 9	$ .27 x 3	$36.58 x 9	$3.75 x 7

Brainwork! How much would it cost to suit up nine players with shirts, pants, gloves, cleats, and socks? Clue: The boxed problems will help you calculate the total cost.

Practice Makes Perfect

Fill in the missing numbers on the multiplication chart.

X	4	1		8	0	9		2			3		5
7									84				35
5			50								15		
	8				0								
1								2			3		
10		10								70			
								0					0
8	32		80									88	
11							66						
3													
9		9				81						99	
				48							18		
12							72		144				
4			40										20

Riddle: What has twelve legs and catches flies?

___ ___ ___ ___ ___
5600 1040 8190 560 560

___ ___ ___ ___ ___ ___
1040 3710 1380 2100 8190 2500

___ ___ O ___ ___ O ___ ___
1170 8190 2100 780 1380 1800

___ ___ ___ ___ ___ ___
780 3710 2850 2850 560 8190

___ ___ ___ ___ ___ ___.
600 1920 1250 1380 1920 1250

A 60×30	C 25×50	E 28×20	F 39×30	G 42×50
H 26×40	I 64×30	M 57×50	N 23×60	P 12×50
R 91×90	S 26×30	T 70×80	U 53×70	Y 50×50

A Tongue Twister:

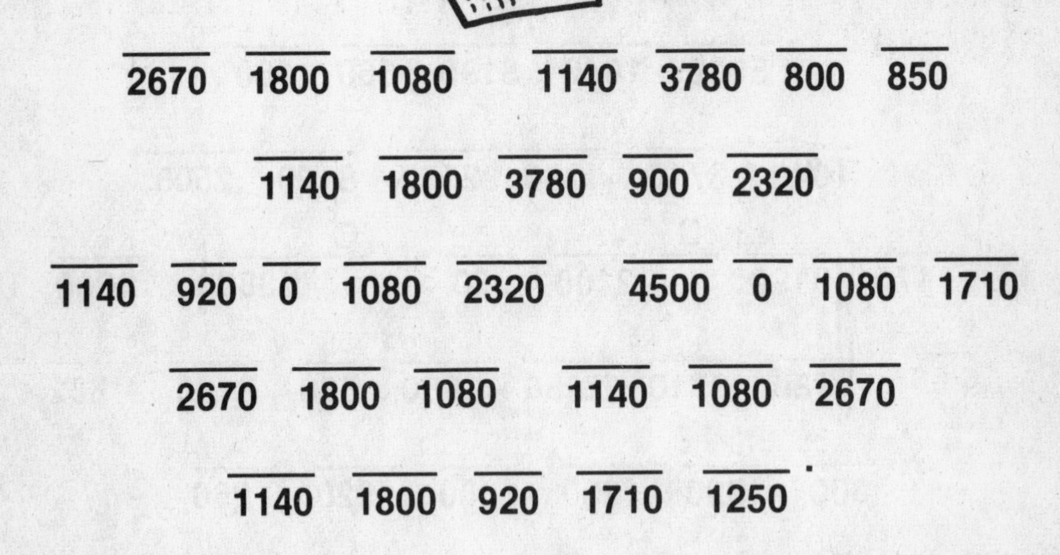

2670 1800 1080 1140 3780 800 850

1140 1800 3780 900 2320

1140 920 0 1080 2320 4500 0 1080 1710

2670 1800 1080 1140 1080 2670

1140 1800 920 1710 1250.

A 23×40	D 17×50	E 36×30	F 25×50	H 90×20
I 63×60	L 29×70	N 10×80	O 75×60	P 45×20
R 57×30	S 58×40	T 89×30	V 90×0	W 57×20

Riddle: How can you get into a haunted house at night?

___ ___ ___ ___   ___ ___ ___
198 156 132 882   144 320 132

___ ___ ___ ___   ___ ___ ___ ___
384 198 198 154   100 860 144 320

___   ___ ___ ___ ___ ___ ___ ___ ___
286   140 273 132 150 132 144 198 882

___ ___ ___ .
273 132 473

A 22×13	D 12×32	E 11×12	H 32×10	I 43×20
K 13×21	L 15×10	N 42×21	O 18×11	P 13×12
R 11×14	S 10×14	T 12×12	W 10×10	Y 43×11

Double Jumping

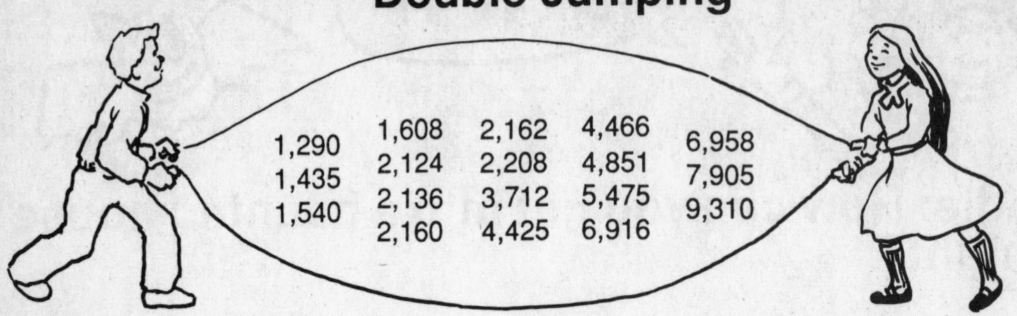

Solve each problem below. Then cross off its product between the ropes.

A. 49×99 | 35×41 | 55×28 | 24×89 | 46×47 | 64×58

B. 75×59 | 86×15 | 95×98 | 76×91 | 45×48 | 75×73

C. 59×36 | 67×24 | 48×46 | 77×58 | 98×71 | 93×85

There She Blows!

Solve by multiplying. Cross off the answer on the whale.

```
  67
x 32
 134
2010
2,144
```

A. 31 x 23

B. 90 x 58

C. 53 x 20

D. 71 x 46

E. 52 x 44

F. 82 x 64

G. 68 x 70

H. 94 x 42

I. 73 x 25

J. 59 x 66

K. 25 x 18

L. 41 x 96

M. 78 x 65

N. 18 x 78

O. 99 x 99

450	1,825	3,894	5,070
713	~~2,144~~	3,936	5,220
1,060	2,288	3,948	5,248
1,404	3,266	4,760	9,801

Brainwork! Solve these problems: 12 x 43 and 21 x 34. Which product is greater?

A Happy Thought:

___   ___ ___ ___ ___
578   625 578 506 437

___ ___ ___ ___ ___ ___   ___ ___ ___
345 312 750 578 350 399   264 612 546

___ ___ ___   ___
300 612 1152   312

___ ___ ___ ___ ___ ___ .
300 1152 578 437 350 144

A 13×24	D 12×12	E 19×23	F 25×12	G 19×21
H 15×23	I 17×34	K 46×11	L 25×25	N 14×25
O 18×34	R 36×32	U 42×13	V 75×10	Y 24×11

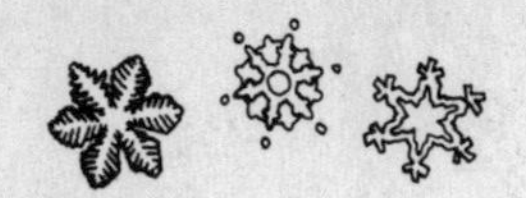

Factor Snowballs

Fill in the missing factors
for each set of stacked snowballs.

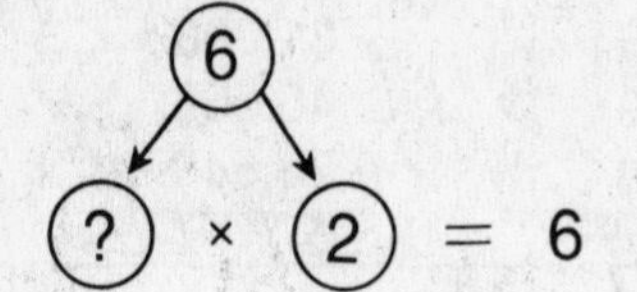

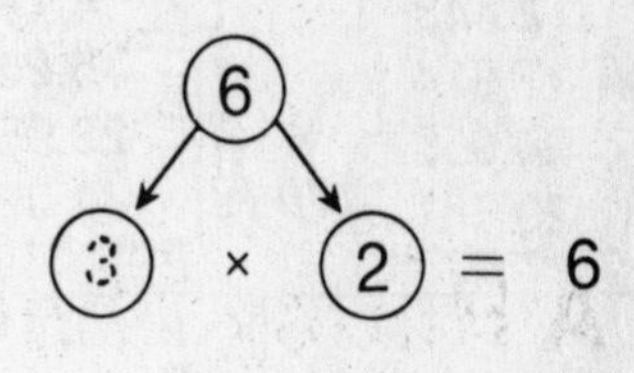

1. 15; 3

2. 

3. 

4. 

5. 

6. 

7. 45; 5

8. 

9. 

10. 

11. 

12. 

54

6 × 9 = 54

2 × 3 × 3 × 3 = 54

13. 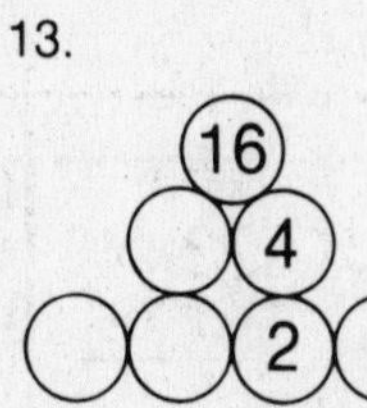

14. 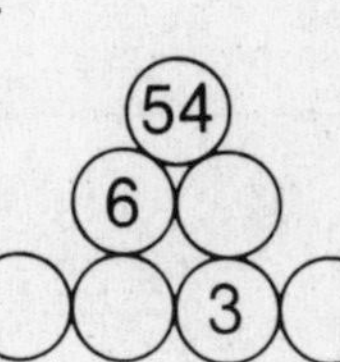

15. 48; 8; 2

16. 64; 2 4

17. 48; 12; 6

18. 36; 6; 3

19. 16; 2 2

20. 36; 4; 2

21. 72; 9; 3

22. 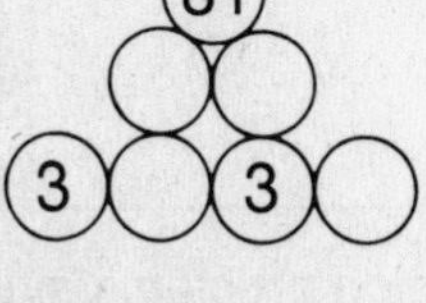

23. 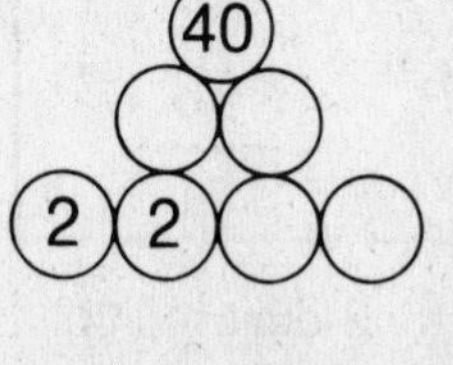

24. 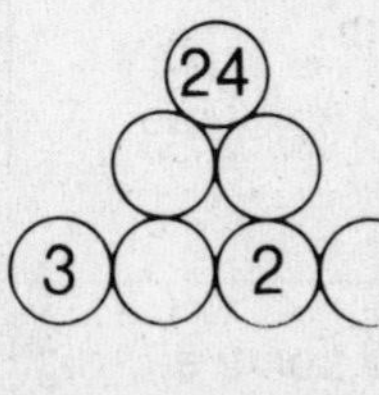

25. 32; 4; 2

Math Robots

Solve each problem. Cross out the answer on the robot above it.

Robot 1	Robot 2	Robot 3	Robot 4	Robot 5
12,443	4,144	15,939	13,266	15,288
17,514	5,236	31,008	18,647	19,200
24,055	23,959	45,568	21,632	41,470
29,681	42,328	78,540	32,652	72,990

A. 834 x 21	**B.** 476 x 11	**C.** 912 x 34	**D.** 603 x 22	**E.** 811 x 90
F. 541 x 23	**G.** 814 x 52	**H.** 759 x 21	**I.** 643 x 29	**J.** 715 x 58
K. 443 x 67	**L.** 148 x 28	**M.** 924 x 85	**N.** 338 x 64	**O.** 196 x 78
P. 283 x 85	**Q.** 247 x 97	**R.** 712 x 64	**S.** 907 x 36	**T.** 480 x 40

Brainwork! Write and solve a 3-digit by 2-digit multiplication problem using 9, 5, 4, 6, and 7.

Homework Helper Record

Color the bean for each page you complete.

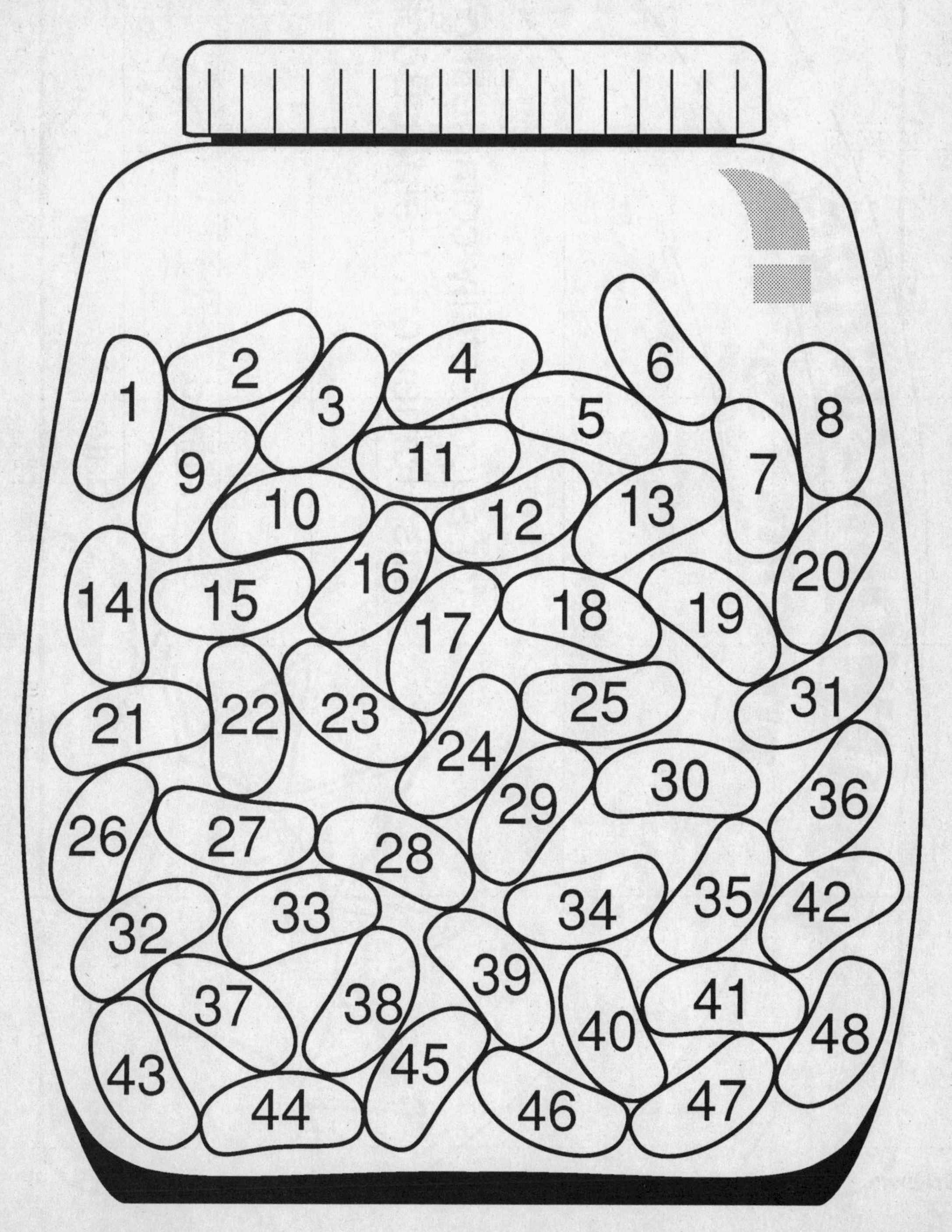

HOMEWORK AWARD
presented to
for successfully completing
this Homework Helper Book
signed
date